Talks With
My Heavenly Mother

Book 1

_____

Ryxi
Renee Shaw

Talks with My Heavenly Mother
Copyright © 2013 by (Renee Shaw)

Blog info:
Ryxi.org

Contact info:
renee.ryxi@hotmail.com

ISBN #   978-0-9795575-3-8
Empowerment Press, a DBA of The Open Mind Foundation

Editing by: James C. Torgersen, Candice Acree Mallicoat

Photography by: Nyk Fry

Cover art by: Ryxi

Cover concept by: Heavenly Mother and Ryxi
Note: I, Ryxi created the watercolor artwork and Heavenly Mother suggested the multimedia concept of Photography within the water color trees. She wanted me to be coming away from the trees with a message. Also, She suggested I wear a crown and carry a sword. At first I thought it would look cheesy, but tried the sword anyway.

Proceeds from this book will be used as Heavenly Mother sees fit. Most likely to help spread Her word. Thank you for purchasing this book and furthering the work of Her new Millennium. If You are being called by Spirit to donate to the cause of Heavenly Mother awareness or to assist with the Earth Grid work, Please contact me at renee.ryxi@hotmail.com.

Dedication

To my journey companion and best friend, James "Torg" Torgersen for the many hours of joint communication with Heavenly Mother and others,verifying information, scribing, and supporting me through the ups and downs of this journey. Unfortunately, Torg passed away a month after the completion of the first version of this book. Even to the very end as he lay in a hospital bed, he declared the truthfulness of our conversations with Heavenly Mother and of Her existence. Without Torg this book would not exist.

I dedicate this book to three of my favorite moms, my Heavenly Mother, the mother that gave me birth and Mother Earth Gaia, After all, I am a Mama's Girl.

First I give gratitude to my Heavenly Mother for making me laugh, spanking my butt and teaching me how this universe works! I thank Her for Her patience as I asked a million questions like little kids do. I imagine it's been funny and annoying for Her to answer all my questions similar to when moms get asked, "Where do babies come from?"

I honor Heavenly Mother and Heavenly Father for releasing my intelligence from the void, creating my spirit and merging the two in this earthly body; for creating a trinity of mind, body and spirit.

I thank my earth Mom who carried me in her womb and taught me unconditional love and the true meaning of service to my fellow man.

And of course, thank you Mother Earth Gaia for teaching me how to heal you and be a part of your re-birthing. For the patience with all of us baby Gods when we spill our milk and poison you. Also, I thank you for growing nutritious food for our sustenance.

This book also would not be here if it weren't for my guides in otherworldly dimensions who paved the way and taught me how to communicate with the unseen world, heal the earth and teach the mysteries.

I feel so blessed,
Ryxi
Renee Shaw

Table of Contents

Foreword

My journey with Ryxi "began" about a decade ago, when I answered a call to help her repair the home she had bought as a "fixer upper". I have been on an interesting journey ever since. She is a fascinating blend, to say the least. Together we've communicated with Heavenly Mother, Gaia and others, traveled to sacred sites healing the Earth together and through it all were best "friends without benefits."

And, lest anyone think this is merely an attempt by her to give vent to inner creativity or to use this book as an attempt to promote her own peculiar world view, let me assure you: NOT SO. She did not venture lightly on this stage, rather had to dragged screaming and kicking. And not always in a metaphoric sense. If I had a nickel for every time she has cried, denied, stormed out in frustration . . . well I would have a lot of nickels! Very little of these conversations fit neatly into her neat, tidy, conservative religious world. Her previous box was so repeatedly destroyed, there is little of it left; mostly just mulch to feed Gaia's gardens.

Make no mistake, this is a book for Christians, and Jesus and His Heavenly Family are featured prominently, but the message is for everyone who has wondered why things are the way they are.

Although she is full of fun, do not take this lightly, or expect to be comforted, unless you expect to acquire an abiding love for the most beautiful Lady you have ever met. And you have met Her before, and She is worth knowing again. The ultimate comfort She is so willing to provide in these coming times is SO worth the trip, you will not look back.

And do not take my, or our, word for it. Any of it. Do as we did, and ask the spirit. Which I fervently pray will be with you always.

In The Light,
James C Torgersen, PhD
"Torg"

Preface

One day while on my morning walk, as I passed through a grove of trees, I heard the words, "You work for Me now." When I asked who was speaking, the reply was, "Heavenly Mother". Tingles, or God bumps, flooded my body as the reality set in. She had popped in randomly before but this was different. I was so excited, I wanted to write about Her and share with the world that She exists! Later I asked, "Can I write a book about you?" She said, "Yes, it is your destiny. Will you tell My story? The next millennium is My time. Will you help so Satan doesn't get hold of it and twist it?"

I've been communicating consistently with the unseen world since before 2005. I've always been curious about ghosts and the spirit world. When other people talked about how they would be so afraid if they saw a ghost, I would say "bring it on!"

I first started communicating with my guides/angels, and, though I didn't know who they were specifically, they seemed to be on track. I connected with loved ones who had passed on, while researching my genealogy. I had this great idea that I could talk to ancestors and get names quicker! Ancestors showed up and I got a few names that later checked out, but they soon told me it was the journey rather than the destination that mattered; that I would soon get to know my ancestors on a personal level, not just a name. Cool!

One day someone new came into the conversation and asked us to "Save Gaia, save the people". What did this mean? How could I do that? I had never heard of Gaia! I am an average person, a single Mom. My good friend, James Torgersen, known as Torg, worked hand in hand with me to verify information and carry out instruction on saving the earth.

They started calling me Ryxi which they said meant "Warrior Scribe". I didn't understand at the time, because I had never written a book before. They said I was a writer in Heaven before I came here and my name was Yashira, which also means "Warrior Scribe". At the time, I hadn't written anything accept journal entries that friends would ask to read. Now here I am writing this book!

Every once in a random while, someone calling Herself "Heavenly Mother" would pop in. Then in 2010 She began showing up on a very consistent basis and now I am writing Her story! I'm a Mama's girl even with my Earth mom so I was very excited to do the job.

How do I know I am really talking to Heavenly Mother? I wonder that myself sometimes. Besides clarifying with Torg, I also will have friends tell me they see Her around me or they will have a dream and tell me about it where the information is the same as what I get.

Many times I ask Her what to give friends and family for birthday or Christmas gifts and She tells me. When I give them the gifts they say, "How did you know?" I've had so many confirmations there is no way I can deny it.

Other times we'll get actual objects come to us from other countries sent to us by a courier! Can't be more real than that!

How do I receive messages? First, I always start my sessions with prayer. Most of the time messages come to me one word at a time like text messages. Especially when I work with Torg.

Other times I get a feeling, see a vision, or sometimes it's like She is right next to me giggling.

At times I misunderstand or forget to write every word, so some clarification might be missing. Sometimes I'll misplace pages or not write down the dates and later try to recall the information. Or I'll forget I already asked a question and ask it twice. At times I even get different answers because things are always changing. In fact, this is a revised addition after she asked me to clarify a few things from the first version based on the reactions of it's readers. She has now given Her stamp of approval so I trust it must be good enough! Or unless she asks me to change it again! Crossing my fingers!

My birth given name is Glenda "Renee" Shaw and my destiny name is Ryxi. I hope you enjoy the ride!

Ryxi

Introduction

Heavenly Mother came to me as a funny, tell it like it is, silly at times and not afraid to be Herself lady. Definitely not what I expected. She says she comes to each of Her children in a way they can relate to Her. In fact, She asked me to revise the first book a little because people misunderstood Her humor.

If you are ready to have your world rocked, your paradigms busted, your box blown apart, this book is for you. I know my world rocked! This book is for and will surprise Christians, Pagans, New Agers, scientists, and alien enthusiasts! In fact I have no idea what category or genre to place it in! All I can say is my own mind has been blown and I hope you answer the call of your own curiosity of what 's in this vast universe we live in and how it all works!!!

There will be answers to questions you will love and totally resonate with and others you may be left scratching your head. I guarantee you will laugh, you may cry and even get down right angry at times. In fact there were some answers I still haven't come to terms with that I left out of the book! You may want to put the book away for a while as you process information. I did. You may love every minute of it and stay up all night reading it. I hope so.

Don't just take my word for it, walk to your local park or go out in nature, meditate or pray and ask Her for yourself if she exists! Better yet, go on a vision quest for a few days in the wilderness and get messages and visions that will blow your mind!

So buckle your seat belts and get ready to blast into the new millennium...Her millennium.

I decided to keep this book as genuine as possible by leaving most of it as a journal of questions and answers. When I first started talking to Mom, I had a million questions I asked one right after another, so the content will be random. Many times I asked follow-up questions later.

Because the questions I asked were so random and hard to fit into chapters based on themes, I did my best to create chapters where either there was an important message or had a lot of information about a specific subject.

Chapter 1

Meeting Her

Note: I will narrate with regular type. My questions will be in bold type and Heavenly Mother's answers will be in italics. If anyone is speaking besides the two of us I will start the sentence with the name of the person who is speaking.

Sept. 15, 2010

While taking a walk at my local park, as I walked the tree filled path I thought to myself, " Where is Jesus?". I realized I hadn't felt Him with me for a while. He had been my guide and friend for so long, had I offended Him? Was I still on "The path"? As I walked through a grove of trees, I heard a voice saying...

You work for me now?

Who is this?

Heavenly Mother.

I got the tingles all over my body, which I take as a confirmation of truth. I thought, "That explains it! What does working for Heavenly Mother look like?". I was excited because I had always believed She existed and had wondered where she was?

Later that day as my Shaman partner, Torg, and I settled into our routine of communicating with the unseen world, we wondered who would show up this time? Would it be my guide Milni or his guide Pongust? Would it be our ancestors, or Mother Earth, Gaia? Every session was an adventure! After our opening prayer, the words came.....

Is Ryxi Chic on board?

What?

Your walk today, you serve me now?

Who's here?

Mom and Gaia are here. Wow, well done my faithful servants!

Mom?

Heavenly Mother.

I felt a surge of excitement at the reality that Heavenly Mother was back, and I could ask more questions! It was interesting to me that She spoke in biblical terms. And, what had we done well?

Why can I hear you in the trees?
They act like antennae.

I thought of all the people I know that say they go to the trees for answers; maybe it's not the trees talking, but God.

What do I do now?
Get over your ego. Not over, be accurate. False humility serves no one. Have confidence in your mission. We four are one. (Mom, Dad, me and Torg)
Do I have an ego?
I say the same thing to all my servants. A general guidance. You are too humble! You need to find a space between pride and standing in your power.
How can I be one with You better?
Pray, dance, sing and obey with knowledge. Many paths. You do the work for balance.

Could it really be as simple as dancing and singing? Our guides had said the same thing. I could see praying. . . and what does "obeying with knowledge" mean? So many questions were spinning around in my head. In the past She had randomly popped in with a quick comment and then be gone, leaving us wanting more. This was different. When She spoke She used choppy words and most of the time incomplete sentences. Like short hand or texting. I will fill in the blanks a little so you can understand Her better.

Have faith my children, We know what We do.

I felt Her power and imagined how frustrated She/They must be when we don't trust THEM. Of course, They have been around a long time!
You did well with your kids, none are lost. Look at Me, I lost many, I knew what I was getting into.

My heart sank as I thought about how She must have felt watching Her children through the ages commit atrocities and how some of Her kids chose Satan's way instead of Hers. She raised billions, I only raised two kids, my odds were better.

Did She mean that She knew She was getting into a lot of heartache over lost children?

I want to serve you full time and have the funds to do it!
Have faith, eternity is a long time.

I was hoping for some specific advice on how I could create a more stable living while I work for Her. Does that mean I would be living moment to moment wondering how the bills would be paid for ETERNITY?! I was a little discouraged.

Torg and I have been doing our calling to help with the Earth shift for quite a while now without 9 to 5 jobs. It has been stressful wondering how the bills would get paid, but they always do!!! Usually at the last minute.

I kept thinking of Her for the next few weeks. The idea came into my mind to write a book about Her. I mentioned my idea to Torg and that I would love to have an appointment with Her once a week with questions for the book. The next time we talked to our guides they said Heavenly Mother wanted to meet me on Sundays, preferably in the evening! WOW! I couldn't wait for the next Sunday!

Oct. 15, 2010

Hi daughter,
Can I write a book about you?
Yes, it is your calling. Only truth. You still work for Father and Jesus. You just tell My story.
We don't know much about You.
Write that it is now time for balance. Write only truth. No ax to grind and don't add or take away. Father and I are one. You and Torg are one.

I realized she mentioned "no ax to grind" because I have been really irritated that the world has been out of balance in the male

11

direction for what seemed like the beginning of time! I'm sure She knew I wanted to right that wrong, after all She is my Mother!

Do you want me to write a book or is it just me?

Yes, both. You need passion to write. It is your calling and destiny.

I am passionate because you've been absent.

Father and I have different callings. Father administers. My 1,000 years starts now. My time is about healing, relationships and family, including ancestors. There have been many false starts, but it was not time yet. You are given a Goddess team now.

She was talking about a handful of women we had recently met who are healers, connected to the earth and open to the metaphysical world. They were supportive of our mission to prepare the earth for the transformation into the new millennium and our conversations with the unseen world and especially with Her.

Will you teach my good news? Go forth....Father, Son, the Holy Ghost and I are a team. All lead, all serve, all have the same mind. When you pray to one, you pray to all. Please pray to the family of Gods.

I liked that. I was glad to know They are a team and that They all watch over us. I felt comforted that even though I hadn't prayed much to Her specifically, that She still answered my prayers. I pictured Them tag teaming prayers..."here you take that one, I'll take this one".

Who is the Holy Ghost?

The Holy Ghost has been many people throughout time and is neither Male or Female. They help Gods teach truth, you listen!

Why do we refer to the Holy Ghost as being Male?

The mission has been mostly Male so far. The new mission is Me, Female. They are spirits, not naked intelligences. They are a whole race not made by Us. There isn't just one Holy Ghost. There is a Holy Ghost always in residence. They are around you and assigned to you and also come when asked for knowledge. They came from the void in empathy for the pain felt here.

Wow, that was a new concept! More than one Holy Ghost?!! Were they like guardian angels? I would learn much more later.

There are lots of people claiming to receive inspiration from God and writing about their experiences. I thought about the prophets in the bible and figures from other religions and cultures who claim to hear the voice of God. In fact, I had personally been guided to ancient records claiming to have been written by prophets and translated by the Mayans, called the "Mentina archives".

You have many people claiming to receive information from you.

Yes, at first many are humble and do it right. They listen and do. Many forget to listen and have their own agenda and ego. Always ask Us.

I guess we are all human, I'm sure I forget to listen, too.

Me too. The tales I could tell! I healed.

I loved that She wasn't always perfect! And I loved Her sense of humor! Maybe there's hope for all of us! I was intrigued and wanted to hear about Her tales! I also found it interesting that She used the word "heal" instead of "repent".

Are you perfect now?

No comment. I can still spank!

So does that mean She is still growing? I thought of how we as parents spank our kids for doing things we did when we were little.

I have spanked my kids for doing the same things I know I have done too.

Good Moms do. Love lots later. I Love you. Good job. Becoming Gods is a process, not an event.

I was excited to know that we are always progressing and that it is the journey more than the destination that is important.

I thought this would be a great opportunity to ask God how I was doing on my journey of progression.

What am I good at and what do I need to work on?
Tall order! Ha ha. Good at asking the spirit, but not so good at acting on it. Peer pressure keeps you from acting on the spirit.

I was surprised to hear I didn't follow the spirit enough. It seemed I was always following the spirit, or at least trying to. I'm not as good at following commandments perfectly though, I like following the spirit of the law better.

Yes, I could see why She would say I have peer pressure issues, I still feel nervous about sharing my experiences in the unseen worlds. Maybe people will think I'm crazy. I still share in spite of my fears and some people love it and some think I'm crazy! I guess She wants me to do it more? I was starting to wonder what I was getting myself into!

I think I'm good at listening compared to most.
Don't compare yourself to others, compare yourself to yourself!
On a scale from 1-10, how good am I at listening and doing?
Asking a 9 and doing an 8. You work on getting healed. You know what to do.

I can live with an 8.

Chapter 2

Sex in Heaven

Oct. 24, 2010

Is there romance where you are?
It is the essence of where I am! Still kissing and hugging. Yes, there is sex in heaven!

She read my mind.

Good, I've heard some people who have had near death experiences say there is no sex in heaven.
Wrong!!!!

I was very relieved, although I am single right now I would hate to miss out on sex for eternity!

How long have You and Father been together?
Billions of years. Still in Love.

I couldn't imagine a couple being together for that long! Had they always been together? Were they ever married to other people? Had they ever been divorced?

Did you ever have other partners before each other?
You would need to grow more to understand. Humans need to grow more.
How did You and Father meet?
He was a flying teacher and he taught me to soar.
He flew a flying machine with an engine and we soared.

I wondered if it looked like any of our flying machines.

Was it love at first sight?
I thought He was slow to ask me to be His.
I should find someone to kiss and hug.
Yes, among other things.

15

How do you feel about the "free love" attitude and that everyone's family so it doesn't matter whose kids are whose?
Wrong. Lots of reason why wrong.

I asked this question because Torg and I meet a lot of out-of-the-box people who have different ways of being out-of-the-box and many have the "free love" philosophy.

When should someone get a divorce?
Go with the spirit.

Some people believe Jesus' mother Mary is Heavenly Mother. Who was she?
Mary gave Jesus a body, We live here. We love her. She's Holy. That's why she came there for that device. Many earth women have trouble with this. It's not about sex.

Nov. 7, 2010

What does a day in the life of Heavenly Mother look like?
Lots. Many people like you need answers, more like nudges. Life on many worlds to keep on track. No night, no sleep. Sleep is a mortal thing.

How do you keep life on all the different worlds/planets on track?
Talk to life.

Do you tell people to make changes?
Works, but other ways work better! I tell life how to grow. This round is almost over. No more new worlds. My round 20 billion years. We came then. You call it the "big bang", silly name. We told laws and watched. Father and I still have fun.

What do you do for fun?
Watch life.

What are your favorite foods?

No need to eat when resurrected. We eat anything we want. None from there. I've never eaten food from there. I think I might like chocolate.

Of course everyone likes chocolate! Isn't there a culture somewhere that believes chocolate is the food of the Gods?

What is your favorite song? (She took a while, seemed to not know how to answer)
From there?
What about Hymns?
Most. Fun, happy, positive. No romantic music from there; your romantic music is about short lives, Ours is eternal.

She started dancing the infinity sign.

I am surprised. What about movies?
Movies with love, adventure and heroes.
What is your home like?
Lovely, Torg saw it in a vision. Crystal and glass.

Torg had mentioned he saw a vision of a castle made of crystal the week before.

Where do you live?
All over the universe. Our home planet is big, in a solar system far away from here.
Why so far away from us?
Weeds.

Does she see us as weeds, I wondered?

Are there plants on your planet?
Yes.
Whose idea was it to start this universe and create the earth?
It has always been the plan. I start when I am ready. The male role is to make the Earth, and female role is to create life on the Earth.
Did you create other planets like Earth?

There are many others in this universe and have humans. The animals vary, I had creative fun!

I imagined her creating zebras and painting the stripes! Giraffes would be fun too!

How did you breathe life into bodies?

You don't understand. DNA. Metaphorical.

Did you create life after Christ made the earth?

Things happen like a car. Father gives laws, I make them sing. The Elements are a part of the team. Levels of co-operation. We say "Make mountains" and they do. We say, "make Ley lines" and they do. We say "make the ocean" and they do. Singing is life. Singing directs life to form. (She sang, "sing, sing a song".)

Laws are dead, worthless alone. The purpose of laws are to create a Matrix for laws to be.

Father drums, I sing. Literal singing. More incomprehensible. I can't give laws. He can't create life. Takes two to tango. He's rhythm and I am blues.

Christ had our help. He did what we did. Lots of spirits acting together as one. We do the same now! You Goddesses heal with just a few. Imagine Billions!!!

Thought becomes reality. We didn't create the universe, you did. We taught you how to do it. Wouldn't have happened without Us. We created the big bang. Everything from then until now We taught and gave callings.

Are there other universes?

Many Gods, many universes. We came from an older model like this. This model works.

I thought about how we humans choose house plans. Did They go to other universes like we go to home shows to see which model They wanted to use?

What makes this universe work?

The only model that ever worked.

Have other Gods tried other ways?

Didn't work.

What are the rules?

Physics, chemistry, and evolution.

How do you create a new universe?

Tell rules, watch. The intelligences were there before and without awareness.

How many other Earths (planets) have you created?

More than you can comprehend.

From where you are, do they look like a pin head?

Less. The universe is big, all things seem small. That's why making Gods is so important.

I wondered about other dimensions here on this earth plane because we have a frequent visitor named Kyle who claims to be a Leprechaun. I used to think stories of Leprechauns were just mythical until he showed up. I wanted Mom's take on it.

Are there other parallel dimensions like Elves, Leprechauns, fairies, aliens?

Some are the creation of other Gods. Some are magical. Some have easy access to here.

Is it OK that we access the others and communicate with them?

Of course. Knowledge is good if you use it wisely. Satan is also in other places. The creatures can be good or bad just like with humans. Some of these are his dimensions, like Kobalds, gremlins, etc.

My guides keep telling me I will soon be meeting little people, is that true?

Yes, soon you'll met them. No hurry. They are here and live in holes.

Were you human once?

As you now are, We once were. As We now are, you should be.

Did you come from a planet like earth before you became a God?

Yes.

What was your favorite memory of your earth?

Two suns and a ring of moons. It was WOW! My family too.

Do you have a mother?

My Earth mom didn't become a God. I still go to see her. My father chose to be an angel.

Could he change his mind?

Not now. Once beings are gone, that's all. Need to have a partner to be a God that creates universes. There were no more women left.

Do you still have a relationship with the God of the universe you grew up in?

Yes, and my physical earth dad too.

You automatically connected your God as Father, what about your Heavenly Mother?

Yes, I grew up in my Dad's time. Think of your earth 900 years ago.

As far as I know, we were not technologically advanced 900 years ago. You must have been more technologically advanced at that time than us because you met Dad flying...

Yes. It was hard to be me, made Me the bitch that I am.

Did Heavenly Mother just call herself a "bitch"? I thought it was funny at first and at the same time I wondered, would a Divine Being swear? Was this really Her talking? I thought for a moment I would delete this from the book but decided to keep it authentic.

What is your definition of "bitch"?

To stand in power loudly.

Don't you tell us we can't swear?

Are you wise enough to know when?

Some are going to like that you swear and some are going to be freaked out!

Good, no pedestal.

For who?

For them. I like Mine.

What was hard about your earth life?

Religious oppression. Technology had nothing to do with spirituality.

So the technology was high and spirituality low?

Much like you.

Do you think we are spiritually oppressed.

Not yet.

I met the wife of a New World Order heavy player who said they were trying to make a one world religion!!! Are they planning to oppress us?

Yes.

Do you think they will succeed?

Try not to allow it. You're ahead of the game.

But what about other countries? We are in a free country.

Internet to the rescue!

There are still people with no internet.

Fewer all the time.

What was your calling on your earth?

Bitch. No, yes, maybe. I made toys. Santa bitch. (stand in power loudly)

Were you a factory worker?

No, made my own.

What did you do for your God? What was your calling?

Teach anyone who would listen. Sound familiar?

Did you stand against religious oppression?

No, didn't know any better.

Do you know how it all started, before you?

Never any beginning or ending. Many mansions, like My Son said.

Did Eve eat the fruit on purpose?

Yes, first daughter from her planet here. Same as on other planets. The fruit was actual fruit. Their (Adam and Eve's) parents are on other planets, we brought them through what you call a "worm hole". Both bodies were transfigured like Enoch's Zion people. She was not actually made from Adam's rib. A metaphor, had to do with bonding.

Who is she?

She had the calling before she was born. She volunteered. She had a perfect body.

What is she doing now?

> *A God helping here. All my children are still here. Many of the Patriarchs and Matriarchs are all over. But they live here.*

Like other universes and dimensions?

> *Just like Ryxi.*

I go out of body a lot and they tell me I visit other universes, I just don't remember very much.

Is Adam a leader of other worlds?

> *All you need to know is Jesus was the savior for all worlds (planets) in this universe and didn't have to die on others.*

That was good to know. I would have hated to see him suffer and die over and over again!

Do we attract the b-day (astrological sign) we are like?

> *Great flow dynamics. Predetermined arrival. The stars and planets are also governed by great flow dynamics. You have it all backwards. Slide tube... I give up, you're not ready.*

I love astrology so was glad to hear about how it works even though "I'm not ready". I felt like a kindergartener whose Mom is trying to teach Algebra.

Chapter 3

Questions, questions!

We are involved with many groups and healing modalities. Do you have any suggestions on healing?

TRY!!!

I was hoping for a magic wand. Did we have any addictions before we came?

No. Potential for. Part of a pre-set.

Do we all have those potentials? Did you have any addictions?

Yes, not over it... Father!

Other than that?

Luckily, no.

What is the best healing technique for addictions?

It's complicated. Some addictions can further one's path. Can be very powerful.

How can an addiction further one's path? I didn't have time to go down that rabbit hole, so decided to put those questions on hold. I assume it's because people who overcome addictions have to become very close to God to do so. As a healer I would really like to help people with addictions more. My heart goes out to them.

When we heal, do our ancestors heal too?

Depends on the problem.

How can gays and lesbians heal if they want to?

Depends on whether they're sick or not. If they're not sick, they are Gods/Angels helping in our kingdom. Not sin, just not GODS that create universes. It takes a Male and Female God working together to create a universe.

That's good to know; I have a lot of friends who are same sex attracted. I would hate to know they are going to hell. I've always known there was more to it than the Bible mentioned.

What about many churches' attitudes toward gays and lesbians?

Community!

I had lots more questions about gays and lesbians but had so many other questions before this session was over, they would have to wait!

How do You place us in families?

You choose. Long time friends. Sometimes we choose.

My own family is really close, but I know of others who are not and have many issues. Some haven't talked to each other in years. It would really be sad to be stuck with people you don't like.

Some families don't like each other.

Yes, they fight. Their agency.

Can we choose new families and be adopted if we can't stand our own?

Yes, but No. We will fix it all later. The concept is still true. You can find new families.

Why is it so important to unite families in the millennium?

The concept is important. It's the essence of the universe. The essence of the release of darkness.

Why do we need more Gods?

To make more universes and release intelligences from loneliness.

What do you think about Wiccans and Pagans?

Many of both groups know more than Christians about some things. They know how things work. Magic is working with nature. They're better at it. God, elements and Nature work together.

This was interesting. The two groups think they are so opposite and yet they really have a lot to offer each other. It is too bad misunderstandings have created such divisions between the two worlds.

What do you think of the different religions?

Each have a slice, yes. Come together, no. We will never join Evil. I call many in many churches. It is better to have some truth than none.

I have felt God's spirit strongly in many places. I've been to Catholic masses, Native American sweat lodges, chanted at Buddhist temples, danced at the Jesus Rock church, chanted Kirtans with the Krishnas, received Deeksha blessings from India, prayed in Mormon temples, and sang with the Hippies. I was glad to hear God is where They are needed regardless if the particular church is perfect.

What about all the different races?

Not yet. We'll talk later.

Not yet? I wanted to know more...

Were there other races before Adam and Eve?

Yes. More information later.

Do we reincarnate?

Sometimes. Most who think they are, aren't. Memories come from engrams. Some have been guardian angels and are confused. The main reason for reincarnation is a long mission.

Torg remembers being a guardian angel for Da Vinci. It makes sense: he is just like him. My guides told me I was a guardian angel for a woman once who was part of a harem in India. Her husband was evil and I protected her a lot. It is one of the reasons I needed to heal my attitude toward men.

Is going to church every Sunday and paying tithing essential for Godhood?

To attend a spiritual community and give of your increase is the commandment.

I feel Like I do that.

Yes, you do this!

I don't give exactly 10% to the church I go to but I give all my extra to serving others and Her. I try to go with the spirit on who and how to serve God's kingdom. I like following the Law of Consecration the way the Nemenhah (an ancient Native American people, the "People of Truth") did it as described in the Mentina Archives. I was glad to hear I was doing enough.

What about money and materialism?

Materials are a trap. Unearned self worth. Can lead to an attitude of superiority.

Is Jesus married?

To Mary Magdalene.

Did they have kids?

Yes, five. All lived to have kids. Two went to France. One born in Wales. Two stayed in Palestine.

Are there many descendants of Christ here on earth?

70% of Europe and Africa are descendants. No natives or Far Easterners are. Thousands.

Should I try to find a husband here?

There are lots up here. No hurry, Zion is big! More women than men, we're all one. Many are called, none do all their calling. You're doing well considering your baggage.

What baggage?

Anger, triggers, single mom, guilt, and Torg. Ha ha.

Was I supposed to marry my ex-husband?

Yes.

I am divorced and always felt bad because I heard a voice telling me not to marry him. We ended up divorcing later and I felt really bad that I didn't listen! So this answer was confusing.

Who told me not to marry him?

Hilti. She was wrong. Trying to be a friend. If you had married a stable white man, you wouldn't be doing your mission. You were supposed to marry him. He's not your bond mate. Your mission is to bring races together. All.

My main guides are Hilti and Milni. All this time I have felt bad. It was good to know I was on the right track after all. My ex-husband has a multi-racial background and my kids are beautiful. Even with my concerns and the fact that we divorced, I always wondered if our path was for a bigger purpose; good to know it was.

Would you call me a Unifier?
Yes!!! Your friend Rider brings nations together and Torg brings Creeds together. You bring colors together.

Lots of others are doing this, right?
Yes, big job. Writing is your tool to bring the races together. The keyboard is mightier than the sword! You were a writer before. Not here.

I knew I was a writer in the spirit world. My guides told me my name was Yashira (which also means warrior scribe) and that I helped fight the War in Heaven with my writing. They also tell me I had at least one past life but won't tell me anything about it.

Was I a writer in a past life?
Satan always feared you.

Cool, I like scaring the crap out of him!
He has no crap! Cain thinks he can beat you! Don't you let him!

Cain? What does he have to do with this? I haven't really been a studier of the bible. Torg explained that Cain had a mortal body, so he knows how to tempt humans from his human experience.

What is my Achilles heel?
Anger triggers.
Anger triggers? Yeah, I have to admit, I have felt like Tyrannosaurus Rex when I am PMSing! My kids tease me and say I am possessed sometimes. I would have to ponder this more.

You will be writing about the races.

No wonder I have always been passionate about inspiring tolerance and have been against racism! Torg and I were so busy and I really was not a writer, at least not in this life yet. I was overwhelmed with the idea of another book to write and fascinated at the same time. I wanted Her to give me a list of priorities to be less overwhelmed and do the most important things first.

I'm spread too thin! What are the most important missions I have?
1st- Bring the races together. 2nd- Write about Me (MOM) 3rd- Prepare Gaia for the shift. The U.S. is one nation with many colors.

I knew about my mission in preparing the earth for the shift We had been getting the earth ready for some time now. These other two missions were a surprise. The book about the races was especially surprising. This should be interesting!

Love you, bye.

Chapter 4
Sacred sites & healing the Earth

There is a valley in Utah where many people have had visions, seen angels etc. One of these people, claims that an angel showed him that Adam and Eve lived and were buried there, after leaving the Garden. He was even shown their burial site. This valley is now called Sanpete County and is located near Central Utah. Our guides had also confirmed this was where Adam and Eve lived. I wanted to ask Mom.

Nov....???? Not sure when.

Did Adam and Eve live in Sanpete County?
> *Yes. Adam and Eve lived there after leaving the Garden. Their kids too.*

Why did they go there?
> *Holy, grid.*

Weren't there other grid places?
> *This one wasn't taken by others.*

What? Wasn't taken by others? I wondered if this meant there were other people in the other grid places at that time, which would mean there were other people on the planet before Adam and Eve? No wonder scientists found very old human remains in Africa they call Lucy.

Is there another grid spot equivalent to this one?
> *Hopi area. (North eastern Arizona, four corners area)*

Interesting...We had visited Hopiland many times and were friends with many Hopi people. I had found an earth energy center there one time while on a hike.

My guides have been teaching Torg and I about the Grid and how to get the earth ready for the change into the new Millennium and for Christ to come back. They taught us how the Earth works so we can do our job.

Why be at a grid spot?
Same work throughout time, Gaia.

They taught us that grid spots (energy centers, portals) are areas on the earth that connect to the Grid (Ley line) system of the Earth. The Ley lines originally looked like the seams on a soccer ball, but with time have shifted out of alignment. There are many different kinds of energy centers on the earth. All feed the Grid system (Ley lines) and most also connect to God.

Whatever happens on these sites affect the grid. Indigenous cultures have been aware of them throughout time. They hold these places as sacred and do ceremonies with singing and dancing for the Earth. When the grid is energized, Gaia can grow our food and have the strength to take care of herself better, like fight pollution. They also do ceremony on these spots to get better in tune with God. Satan's people also try to influence the Grid and do evil ceremonies that bring the grid down and claim the grid for themselves and Satan.

Many churches and Temples of all religions are built on these portals to the Grid so they can connect to God and their songs energize the Grid system. Spiritual leaders who are in tune are guided to these areas. Most do not understand why. The Holy Land is one of these places. Solomon's temple was built on top of a major energy center called a "node". Other node centers are Stonehenge, Pyramids of Giza, Machu Picchu in Peru, Nepal, Alice Springs in Australia and Mesa Verde in Colorado. Now She has confirmed what our guides have taught us that Sanpete County in central Utah is another energy center, although it is different than a node. I'm still learning what the various types of centers are for and how they work.

Was Noah in Sanpete, too, and was the Ark built there?
Yes.

The same gentleman who claimed Adam and Eve lived in Sanpete county also claimed that Noah's Ark was built there too. Torg heard of a scientific study performed on the wood of an ancient ship found at a site in Turkey that many believe in the Ark. The wood samples have been proven to be the same as cedar wood growing in Sanpete County, that is unique only to Sanpete.

What about Adam's temple?
Yes, you go find the temple of eternal joining!!! A Quest! Your guides will help you.
Can you give us any clues?
Could, won't. Plant trees at Eve's temple in the spring?

We had learned a few years earlier there were two temples in Sanpete County, one called "Adam's Temple" and the other "Eve's Temple". We were asked to do a ceremony to join Male and Female in balance for the new Millennium. I was told Adam and Eve had been waiting for this time and I had been called to lead this ceremony before I was born.

I felt inspired to hold it on the autumn equinox. I found out later that this was the exact time the Mayan Calendar claims that Male and Female will come into balance! I was amazed, because sometimes I'm not sure if all I do is real. This was a big confirmation for me.

Many people there had visions and dreams of Adam and Eve along with other prophets being at our ceremony.

We were taught by our guides that Adam's Temple was made of rock, signifying knowledge and stability. People would go there to learn. Eve's Temple had a grove of trees where people would go to sing and dance.

They were across the valley from each other. Our guides explained there was a place in between the temples where people would go to get married, hence: "The Temple of Eternal Joining".

What is the future of Safe Haven Village?
Open, depends on members' agency.
What would You like there?
To fill valley with righteous people.
Anything particular?
You know.

31

The portals?

> *Some. Mystery school. Sacred place.*

What is the history there?

> *Many Holy events and people. You go close sinkhole portal! Needs keys you both have. Your guides will tell you how.*
>
> *Your job is to assist with several cities of light.*

Safe Haven Village is also in Sanpete county. A few years earlier I had been guided to a series of records that had been found and translated that had been hidden is the mountains near Eve's Temple called "The Mentina archives". They were a history of a people called the Nemenhah who lived in the Sanpete County valley after Adam and Eve. Apparently people throughout time were attracted to this place and lived there at different times.

One of the prophets who had written in the records warned to whoever read these records it would be time to find safe places in the mountains. She was told the records would come out at a time when the world would soon be cleansed. I took her words seriously and started looking for a safe place in the mountains. I met others who heard the same call and we ended up buying a property in Sanpete county and calling it "Safe Haven Village". In fact one of our members has felt the call to build cities of light since she was 9 years old.

My guides had asked me to create a mystery school/wisdom center at our community we are building. In the meantime we were asked to teach in our home.

Why is Chaco Canyon a more powerful Earth energy center than other nexuses and even nodes?

> *The net is very complex. It was a place of humans joining which made it more powerful.*

Chaco Canyon is in Northwestern New Mexico and has many ruins similar to Mesa Verde. My friend and I had passed through there and I had found an earth energy center there, too. I was asked by my guides to re-activate it. When I did I was told to make an infinity sign in the snow by walking. There is definitely something important about the infinity sign. Mom dances it and now I was told to make one.

Is the earth over populated?

Plenty of room. There has never been too many people. Humans need to get smarter.

Nov.14, 2010

Torg and I had been invited to listen to a group trying to change our government back to its original state. They were trying to create a new government because they felt ours is corrupt. I agreed that our government is corrupt so wanted to see what they were planning to do about it.

What did you think of this weekend with the New Republic?

Loved watching kids play, so serious. The big picture is right. Is time for a change. Was funny. You ever watch your kids?

What was funny to you?

Running out of things like paper, staples, food. They were so serious!

What did you like?

Prayers and sincerity. They talked too much about nothing. Your government will survive. I've made it that way. No need to change to a new government. Fix, yes. Educate, yes.

Whose job is it to educate, us?

Good guess! What is your calling, Ryxi?

Warrior Scribe?

Imagine that, Ryxi! Not all My ways are mysterious! You did very well the last 3 days. Taught many mysteries. You reached many people.

The New Republic recorded the event and had a live streaming video running where many people around the world joined in. I commented many times.

How's my school doing, by the way?

She was talking about the mystery school our guides told us we were supposed to start. So far we were only teaching in our home.

You're healing, getting younger.

What were your other favorites for the weekend?

Ryxi Prayer. (I prayed at the medicine wheel and in a circle prayer and sang.)

Torg's lecture. (He lectured about Common Law.)

More people now heal. Many people coming. Soon you'll be very busy!

Torg fat, fix. Ryxi too sexy, adjust it.

Too sexy? I'm not trying to be.

You have lots of Pheromones. Matrix is healing you. Out of balance. Torg is immune to you, shielded.

One of our new friends our guides call "Rider", also one of the women She refers to as one of "The Goddesses" was teaching us a new healing technique called "Matrix energetics". I guess it was working too well!

Buy a new dress. Something Mom like. Not too sexy. Not too short. No breasts. Cleavage OK, not too much!

You are so funny tonight!

Good mood from watching you. So funny. Fire and Brimstone next week!

Any advice?

Dance, sing, pray.

Anything new?

Same yesterday, today and tomorrow.

I feel I need Torg to protect me when I go certain places.

Torg would beat up someone to protect you.

34

Torg: Thanks.
> *Still fat!*

What is righteous in your book?
> *Nobody's perfect. Do your best.*

What's the best way to fix this country?
> *Step up! You're the Ryxi!*

Nov. 21, 2010

How are you?
> *Godly.*

How was your week?
> *Time flows. Will you please eat better? Greens, fresh* fruits, fish, etc.

Sorry, I didn't obey the Sabbath today, took my former mother-in-law out to lunch.
> *No worries, you're building My kingdom. Sabbath rules are a guide. Not set in stone.*

I was so glad to hear this!! Some people I know are rigid about this and I am soooo not!

Who lives in the core of the earth?
> *Won't say.*

I assumed this means there are people or some kind of beings living there or She would say.

What was your name on your Earth and do you have a name now?
> *My earth name was Mithra. Yes, I now have a name that is sacred that I do not tell.*

I would love to name a girl Mithra after you.
> *It's not too late!*

Too tired and don't have time or a husband.
> *Hire a man? Ha ha.*

Who do you think would be a good choice? You are all knowing.
> *Who told you that? We know a lot.*

35

Torg's calling is to be out of the box and to hang out with you. Hard! Ha Ha.

Between my PMS and anger issues?

Yes! Do you eat garlic?

Sometimes. I forget to take my herbs for memory.

Where did I put that universe! Ha ha.

When do those who graduate to Godhood get their new "GOD" names?

You'll find out when you are healed. The name creates you. You create the name. Not a light choice. It changes the universe.

Do some people purposely keep themselves from resurrecting? Can they heal and choose not to resurrect?

Yes.

We had a spirit show up to help with a winter solstice event once. We were told she decided to stay here instead of crossing over so she could help here during this time of earth changes.

You are healing your Mother-in-law!

How do I heal her?

Be her friend, daughter. She's scared and lonely. Many miracles. Stand in your power.

How do we stand in our power and still give You the glory?

Glory is over rated. You or I can't stand alone. We are one. My work and glory is you. (Her kids)

Will you be glad to take a break when this round is over?

Too soon to tell.

I'm sorry I haven't done everything you've asked yet.

Of course not. I ask a lot of my best kids.

Chapter 5
The Elements

In 2005 while Torg and I were having a session communicating with ancestors, something or someone new came in and asked if we would save Gaia and save the people. I had never heard of the word "Gaia" before so I looked it up on the internet and it meant "Mother Earth". When I asked who was speaking we got the words, "Elements....Earth, Fire Water and Air." I was shocked that the elements had intelligences and they could actually communicate! That was the beginning of a 5 year journey of learning how the earth worked and how to get it ready for the transformation into the new millennium. They were the first to call me Ryxi, Warrior Scribe. We wrote a book together about the beginning of our journey called,"*To Dance with Elementals*".

There are different levels of power for each of the elements. We were first introduced to our local element beings. There is an earth element who calls herself "Kimjin". She can open doors and move things around. Before we thought we had a ghost doing all these things. We found out when we light candles that fire elements show up. We had a water element from a spring close by that would visit and hang out in our bathroom. We found out they like chocolate, so would leave chocolate for them.

As the years went by we met more powerful elements governing larger areas. We became friends with earth elements that governed local mountains and one that governed the Salt at the Great Salt Lake, who calls herself Nolmi.

I met two elements governing the green granite and selenite mountains in central Utah named Hilti and Milni who turned out to be my guides! They even tracked me to Los Angeles when I lived there for seven years!

Soon, to our surprise, intelligences governing sacred sites around the world were tuning into us! It was amazing!!! Places like Stonehenge, who called himself Pongust, Hawaii called herself Pele, the Pyramids of Giza called himself Pisis, Mount Machu Picchu calls herself Picci and more! Even an element that orbits the earth, calling himself Prescott, checked in! It was very exciting! As you can see most of these powerful beings' names start with P so we started calling them the council of P's. I hope to write an updated book about it some time. Our book, "*To Dance with the Elements*" just scratches the surface!

Dragons are the most powerful of the fire elements. They use the word "dragon" because that's what the druids used to call powerful fire elements. They taught us that in the beginning dragons danced to structure the earth and her Ley line system.

Some dragons would later be offered the option to become human and walk the God path. One time I had a vision of a white dragon dancing. I was told this was Jesus. He took the form of a dragon and then was born with a body later. Torg was a dragon, too, and remembered the original dance. Sounds crazy, huh?

Nov. 28, 2011or 10

Who is Gaia, and why choose her intelligence to be earth?

Same reason I chose you, aptitude. Made offers, either accepted or refused. Most dragons refused the God walk. Dragons were offered their spot. Later were offered human spots. Humans and Dragons have similar aptitudes.

Were you a Dragon first?

No, but Father was. Others also and angels. Two other levels of Celestials. Lots of choices. They like their roles.

What are aptitudes of Dragons?

Make decisions, wise, not homebodies, warriors, they act. Gods do the same.

I think I'm like that.

You're better off with your calling. I have billions of Dragons, very few of you.

What makes me different?

Your intelligence fits this need.

Why Gaia?

Strength, follow through, dependable, etc. All things here share in her spirit. She has no affinity to any element.

Prescott outside (space) has a spirit, gives good council. He is uniquely qualified, just like you. When a naked intelligence shows up, we read, then make offers for callings. The higher on the scale, more choices.

Again, Prescott is the intelligence that orbits the earth and watches. He is the only one with that aptitude. He was promised even though he did not take human form, he would become a God.

I understand, I have so many choices in artistic avenues, it's overwhelming sometimes!

Art isn't your highest self. Communication, is your highest self. The bigger picture...(Art, dance, write, etc.) As long as anger doesn't get in the way. This is a war.

When do we remember the Big Bang and up to now?

When you resurrect.

What do we need to do while here if we remember everything later anyway?

Study, habits you learn there you take with you.

What are the reasons for being here?

Learn to become Gods, make your spirits ascend to their highest potential, commit to Our kingdom and develop relationships. (Family) The level of intelligence created here rises with you. The purpose of life is to find one's self.

You are here to do ceremonies of commitment to Us, make your spirits celestial and develop relationships.(Family)

Do you mean the family as a whole God family or individual families?

All, of course. Relationships run the multiverse.

So not just our blood family.

Too narrow a view.

What is your description of Celestial?

God, creation.

So it's not good enough just to go to church on Sunday?

Hell NO!!!

So this book will blow conservative religions away!

Heavens YES!!!

What would you tell the typical conservative church goer?

Live every minute like a God. Make things happen! Very individual. Just ask God . . . oh that's Me!

You're in a good mood tonight!

Happy because week of gratitude. (Thanksgiving)

Do you like Christmas?

Yes.

What is your favorite Christmas song?

Rudolf. Nose means Holy Ghost. Should have been 12 reindeer. You talk to me without the trees now, their sleepy, it's winter.

What do you look like?

What I want. You can too. White and glorious with nice shoes.

Torg needs a haircut.

Do you make your shoes there?

I ask intelligences (elements) and magic!

So you visualize shoes, ask and the elements show up in the form of shoes? Is that unrighteous dominion?

Not if I ask. Good Ryxi.

What is your take on my friend Kathy?

Lost spiritual anchor. She needs to fix her alcohol habit here.

It's hard to fix people like that.

Can't. She needs to fix it herself or it doesn't get fixed. Self first.

I finally set boundaries with my friend Gwen. I was pretty blunt and firm with her. She took it pretty hard. Was I too hard on her?

Tough love. Was the flood too tough? Hard decision.

Did you have to do that on other worlds?

No, thank Me. Ha Ha.

I've heard people say our earth is the most evil planet in the universe. Is that true?

Yes, there are also more Gods. Much agony here. Graduate more Gods here. Others are more bland. If there are no wars, there are no warriors. No tests, no Gods.

Why do we need Gods again?

Lot's of intelligences want to progress.

Why is it so important?

It's never ending. There will always be more, endless intelligences to save. Do you feel OK about leaving others in darkness? I honor my God for bringing me out when he did. The big bang was mostly about law. Men like to blow things up.

Chapter 6

Women Leading

Dec. 5, 2010

Hi, Mom. Sorry for the Chaos. (We had lots of drama happening around us)

Used to it.

Your kingdom is organized?

Yes, but Earth is not, Yuk.

We're like Weeds?

Not really, tried for a metaphor. Didn't work. I like the book. You took out angry stuff about men being in charge and screwing up. It was hard I know. Yea, it was hard to watch the whole world go to hell before the flood.

Will women do a better job leading your time?

You should read about Cleopatra & Bloody Mary. Women tend to be passive/aggressive. Just as bad. No progress without stress.

What type of women do you want on my Goddess team?

Need lots of variety of women. We need sensual/sexy, young, old, Mother, Grandmother, lesbians, virgins if there are any. No rigid or loose. Righteous women.

Strict religions will be surprised!

Good. Sexy is OK. We condone sex within marriage. The Puritans were so rigid, joyless. Not healthy.

Why don't you change it?

It's not supposed to be this way. It is evolving.

Balanced sexy right?

My people are not promiscuous. Sexy means "joy in sex", not flaunting it. Sex and family attraction should be joyful. Satan knows this and twisted it. He used sacred attraction for lasciviousness. Inappropriate sex. Prostitution. Whatever breaks up families.

Lesbians?

Lesbian families should be sacred too. Not Gods that create universes and spirits, need to have both Male and Female to be Gods.

Maxine, no joy, frigid. Leads to abuse. Both extremes not good. I prefer hippies over rigidity. Time to evolve. You're right. Goddesses are sensual without lasciviousness. You are part of a select group. I am proud of you.

Maxine is a spirit who asked me if she could be my God Mother. She was an herbalist from the 1500's who couldn't have kids and was married to an abusive husband. She had to hide her herbal practice for fear of being called a witch and burned at the stake. She wanted desperately to share her knowledge. She has been visiting me and teaching me about herbs. I started a book to honor her and share her story.

She also came from a Puritan era where sex was considered bad. She was a very crabby spirit. In fact, when she first showed up she acted like an old fashioned school teacher who cracked the whip! She would say in a stern way, "You learn herbs this week, test on Friday!". I agreed to learn herbs if she would agree to have fun that week. Some weeks I would challenge her to sing a song. I even asked my deceased Grandpa and brother to teach her how to have fun. They are pranksters and like to joke around. They reluctantly agreed. Soon she reported that her friends liked the change in her and that they enjoyed her company. Now she has a boyfriend!

**I'd like women from different cultures and races...
Africans, Indians, Asians etc.**

Now you're thinking! I open hearts, you open minds. The next 1,000 years will have new traits, skills, and needs.

Many indigenous cultures have prophecies that say, "When the grandmothers gather, the world will know peace." There is a group of women called "The 13 indigenous grandmothers council" who started gathering in 2004 to fulfill that prophecy. Many of them knew they were called when they were very little.

They are one of my councils. You also were called before you were born. You and your friends. World Peace.

They are from the four continents.

You too. Do you see yourself in prophecies?

No.

You were, you and your group. The Nemenhah prophesied about you now, that women of power would come.

I haven't seen anything like that.

It is, more are coming.

How many in our group?

Thousands. Big movement.

Who are some of the top influential women since the beginning of time?

Mary, mother of Jesus, Mary Magdalene, Ruth, Esther, Eve, Joan of ark, Panatan (Nemenhah records), a queen you don't know and others not in history books.

Why didn't Mother Teresa make the top 20 most influential women?

You people made her famous, not us. Still is great but is not the top 20. She emphasized poverty too much. Set goals you can't do. Went overboard on being poverty oriented. Absolutely no possessions is out of balance. Clara Barton is ahead of her. Margaret Thatcher and Rosa Parks too.

Should I have taken my mother-in-law to dinner on the Sabbath?

I say it's OK. I made the Sabbath, I say it's OK.

Torg still needs a Hair Cut.

Go dance.

Sleep. Practice going out of body.

?????

Why did Paul in the Bible say that women had to obey their husbands?

Not exactly what he said; he needed converts. Needed to be politically correct. At that time men dominated, women were submissive. Archaic, will you fix?

So basically, women had to be treated like shit so you could get converts?

Other female cults at the times were evil. Had to look better to get converts.

So you traded one bad for another bad?

No, we had to look better. They were worse. Prostitutes.

So they were supposed to obey their husband?

Husbands are supposed to be righteous. Wives had no obligation to follow unrighteous husbands.

Seems like a sideways thing.

It was. The times required it. Information wasn't meant for all...a private letter. Paul's letters weren't supposed to be in the Bible. Evil men later wanted women to obey and put it in when they chose the writings for the bible. Wanted to control women. Did. Paul's people were good people, product of the times.

Titus and Corinthians were not supposed to be in the bible.

Why were the women prostitutes, were they poor?

Money and power.

Claiming to be Goddesses?

Some. Vestal. Oracle Delphi. Look it up on an empty stomach. Venus cult and others. Women have been as evil as men. Satan uses all to cause fall.

Why was it OK for men in the Bible, even prophets, to have more than one wife and concubines?

There were more righteous women than men. Most "concubines" were wives. Political reasons to call "concubines". Needed to not appear as a female cult. Had to be opposite.

Dec. 13, 2010

How do you like the book so far?

Great. Want to write about more conversations...like a movie script.

Will you proof read and change anything I miss?

Yes, I'll proof read for you. In sentence 5 the grammar is wrong, Ha Ha.

Why do you use evolution instead of just having babies?

Gods can only have spirit children.

Why start them from scratch?

We need many types of plants and animals.

Have you ever transplanted species from universe to universe?

Humans. It's complicated. Would take too much time. Get a science book. What do I do with a dinosaur intelligence without a dinosaur?

Do you work with the Pope and other religious leaders?

Yes, ALL leaders. They have followers and need help.

So if everyone listened to the spirit, everything would be perfect?

Only if they follow it.

Love you Ryxi Chic, walk conscious.

What do you think of Siddhartha, Buddha?

The fifth most powerful and enlightened of my sons there.

Who are some of the top?

Jesus, Adam, Enoch, Joseph of Egypt, Buddha, Abraham, Seth & Confucius.

I've been so overwhelmed. Everyone seems to need me all at once!

Join us! Imagine 8 billion babies needing us all at once. Disobedient brats!

How's everything on your end?

Doing good. We do Celestial palates! Ha ha. Tomorrow at 2 afternoon.

You served us all day.

Torg nice hair. Good job Ryxi. (I gave him a haircut.)

Torg finish shield for protection. Proud of you. Hugs. Go heal generations. (He had healing appointments)

Rider talked to your Christmas tree. Give her a blessing.

One of our Shaman and Goddess team members is known by our guides as Dragon Rider, Rider for short. She loves to talk to trees. Our guides said the earth rocked when we gave her a blessing! They say when three people work together magic happens!

Dec. 20, 2010

Sorry about last night. I missed our appointment because I was helping someone.

No problem. I like watching my kids do good works.

Sorry about wondering if this is all real.

It's normal to wonder if I am real.

I had been wondering if I was really talking to Her or if it was just my imagination.

Why don't you hear my thoughts?

I can, you have your privacy. Do you go through your kid's rooms? I choose not to listen.

Where do animal spirits come from and where do they go when they resurrect?

They go to the level that is best for them. Resurrected animals create spirits for mortal animals. It's close to the way We create. Like how human spirits are made, except human spirits have to come from Celestialized parents.

What about plants?

Extra spiritual matter that flows from celestial plants to all plants on all planets (called "deva")

Are there more plant spirits coming all the time?

No, there was more there during the plant ages. Earth is losing plant spirit matter now because plants are dying.

When plant spirit matter goes away, where does it go?

Where it's needed. It can even go to other planets.

If we need more, will it come here?

It happens more if you sing and make plant matter grow. Celestial plants will make more spirit plants and they will come.

Do you bring seed stock from other universes?

Haven't yet, maybe? If I want a rose, I evolve one.

Can people travel from different realms, universes, etc?

Celestials can travel. Others can't.

What do people do who don't make it to Godhood? Stay in this universe?

Celestials go where they want. Others stay at the level they ascend to.

What do the ones that stay do?

We brought Celestials with us here in the beginning to help create this universe. Can't leave a level but can visit other planets with help, but it is rare.

Jesus watches over this realm (Earth and other realms that have developed this level) and the Holy Ghost watches over lower evolved realms. This is Jesus' Kingdom. We are the parents.

Yahweh is Jesus. Other universes have chosen Sons to run and save. We started it, made spirits and chose Jesus. We're well pleased!! He rocks!!!

So we can go out like You, start a new universe and choose a Son?

He could, too. But it never happens.

Why not choose a daughter?

Needs Male energy.

He won't have spirit children unless he starts a new universe. He only has mortal children. Posterity too. He will visit. When he created (male stuff) and when I did the female part (create life), He didn't have a wife at the time. He was only a spirit. I gave life.

Jesus was the first God intelligence to get a spirit. Big Bang, all show up.

How did you create life?

Take chemicals, add spirit and Bingo!! Get a book on chemicals. It's magic!

Is that what you would call Women's Priesthood?

Women's priesthood is empathetic and life giving. Male Priesthood is laws and administration, boring! All have a mix. Some men have empathetic symptoms for their wives. It's cold but similar. We have both. Work together in balance.

The Nemenhah women were administrators. Some were even high priests.

Yes. They went through a process, too, and took a long time to get there. The Nemenhah records are the end product.

You and Torg bless together. It will soon be common for men and women to do blessings together. Join with partner. The Millennium is about balance.

Would you like for me to write more about Women's Priesthood?

Yes!!!

What do I write?

You already know. Nemenhah records (Mentinah archives) were written by people to you, but it wasn't the way they really lived their lives. It was the way they were supposed to live.

Love you. I go now, Ryxi Chic

December 26, 2010

At first, Heavenly Mother had a messenger come and say She'd be here soon. He said he could answer simple questions for Her, so I started with the first question on the list I had created earlier.

Do Aliens come to our universe and do we need to protect ourselves?

No, we have enough evil of our own. Yes, they come to learn. Humans from other places come to visit in space ships.

What about green/gray ones?

They are humans adapted for space travel.

What about abductions?

A lot of abduction stories are not true. Space humans have kidnappers like you have kidnappers. You don't have to worry about Aliens from other universes, only some from your own universe. There are good and bad ones. There are systems in place to stop them. They have their agency like you do. What you call Aliens are humans from other Earths that have learned space travel. You too soon will do the same. Some were even Earth humans once.

How long ago?

Before the Flood.

They were that advanced before the flood?

Technology, yes. Morals, no. You're getting ready to do it again! History repeats itself. They must have learned morals or their culture would not have survived. They would have killed themselves off. They are mainly adventurers and scientists. Some are messengers. They want to help like your Peace Corps.

Others were saved from the flood too.

Yes, many cultures survived, not just Gabriel's (Noah's) people. Some went underground like the Hopi. People like Noah made boats and others climbed Mountains. Most that survived listened to God.

Why didn't they keep their technology?

All went back to Gaia. High Tech not needed.

What were some of the worst things they did?

Anything that destroys the family. Sex without marriage, sex with animals, kill children for magic, sacrificed virgin men & women for magic. Deny and defy God. Abortions. The family is the core of the Godhead.

Satan used sex to destroy families and the kingdom. We could not look like evil.

So this has happened a lot.

Through all time. Baal at one time was a statue that roasted human sacrifices- sent by a Goddess. Prostitutes were sent to parents of victims as a reward for sacrificing their child. Satan was behind this. Very slow process. Victims were virgins. False Gods everywhere, every place. Since Michael (Adam) *and Eve came out of garden.*

Others beside Adam and Eve?.

Not on My table. I talk about the pure family.

What is the pure family?

Our kids. Others have other paths. A different Heavenly Mother and Father.

Why are they here?

Renting space. Pueblo and Bush men may join you if they want. Many Bushmen are leaving, no kids. It's complicated. Not the first to go.

I know lots of people who talk about manifesting and creating their lives. How do we balance our will with Yours while creating?

You should be creating with God and Our will in mind. Not your own.

What happened to the Dinosaurs and why are they

extinct?

Most died because of the weather. Their time was over. A big rock hit South America.

Do you feel bad their whole species is gone?

Not their time.

Are their spirits with You?

Some are. Some are resurrected and in their ascended realms.

Will all plants survive to the next millennium?

Pernicious plants won't survive. Only evolved ones will survive.

If we can do anything Christ can do, can we walk on water and other miracles now?

If you have the keys. Why would you want to walk on water? Takes effort. There are better ways to use your time.

Why did He?

To get to the apostles and to build faith.

What is Your favorite Holiday?

Easter, because his mission was done. We always had some question He would succeed.

What do You do to celebrate?

Give hugs.

When do you relieve burdens and when do you let people suffer?

You aren't here to suffer, you are here to learn. I only use suffering if it is the best way to learn.

I'll bet You love all Your kids and there were some You may not like!

Yes, Hitler was a jerk! I had to spank him hard when he got home!

What is he doing now?

Time out.

Is he remorseful?

Not really, bad earth habits.

How am I?

Not bad, I've known better. You're fine. Ha Ha. Abraham is boring. Always talking about his kids.

Isn't that what we are supposed to do.... love our family?

Yeah family, three cheers for Ryxi!

January 9, 2011

When we pray silently, how do You hear us and not read our minds? You mentioned earlier You could read our minds but don't so we can have privacy.

As soon as you say, "Dear Heavenly Mother/Father" we listen.

How do you keep track of everyone?

Good thing I'm a GOD!

I'm glad You have a sense of humor. I'm glad you're not so serious like church can be! It gets boring sometimes!

I fall asleep too!

Can't You do something about that?

Don't you watch your kids perform?

How do you hear everyone's prayers?

Very well. We use the Holy Ghost.

How do you want me to start blessings?

By the power and authority given to me by Heavenly Mother....

Are you OK with the way the Pagans worshiped the "Goddess"?

They didn't worship Me, they worshiped the Earth Goddess. They are confused. I don't mind.

When would You like the book to be finished?

The book will never be finished. Just give what you have at the time. GOD BLOG.

Anything else You want to tell us?

Most of My dresses are white. Sometimes I wear other colors. I get bored with white. Most here wear white. That's why I come and see you! I have some lavender pants.

I would get so bored wearing white all the time too!
> *You're my kid.*

What about shoes, jewelry, etc.?
> *I go barefoot most of the time. No jewelry. No underwear.*

Hair?
> *Any color I want.*

Makeup?
> *Can't improve on perfection!*

Are all the other God's as fun as you?
> *Martha's boring. Don't tell her I said that!*

What are your favorite flowers?
> *1-orchid 2- Iris 3-Lavender 4-Daffodil*

Sometimes I am busy on Sunday, is there another day that would be better?
> *No, just give me a three day notice.*

Jan. 16, 2011

What do you think about Indigo and crystal children?
> *Always have been exceptional. Many more now. Indigo's are smarter, empathetic and spiritual. Not totally changing anything.*

They seem so different than everyone else, they seem to be changing everything.
> *It's always been that way. Bright kids have always been different from their parents. The last time we sent large numbers like this was in the 60's; you call them hippies.*

Why circumcision?
> *Yes, our idea, a mark as special.*

I was hoping she would say it was a cultural mis-understanding. Seems so barbaric. Why not pierce an ear or something?

I have a lot of friends who use Shamanic drugs, and many indigenous cultures use them with great results for physical and spiritual healing. Because we work with people who care about the earth and I am an artist, we keep running into people who smoke pot. They say they need it medicinally as an anti-depressant or pain reliever. It makes sense, but I don't like when they hang around in a group and smoke for fun. Especially when it is against the law. We always end up in arguments over it. I don't want to be judgmental about it.

We've asked the elements about it many times and they say, "It makes Gods stupid." They also say when people use mind altering substances, their shields go down and anything can get in. We always encourage people to get visions and inspiration naturally, but many want us to use these enhancements.

What about natural Shamanic herbs like pot, Ayahuasca, Peyote, or mushrooms?

No, too many of you fall. If people could handle it, it would be OK. A person should ask for personal revelation for themselves. Wine too, sometimes can be medicinal. Most people wouldn't hear the right answer and abuse them. Cigarette's are evil. Kill and addictive. No pot for recreation.

The scriptures don't say to get personal revelation and follow the spirit enough.

ALL scriptures do. You people don't get it. It was written in physical, applied in spirit. Fat people aren't following the spirit. If you follow, you will be healthy. We can't be too specific cause you won't need to listen to the spirit and learn. Smoking cigarettes without ceremony is not good. Alcohol addiction, no. Mind altering drugs, no.

You should treat the animals well.

Since Torg and I are Shamans, and I smoke the peace pipe (we don't inhale) when I run lodges, I wanted to know how She felt about it.

Many Native Americans smoke tobacco for sacrament. In fact when I first started my Shaman path a Hopi Elder asked me to smoke with him. I grew up with the teaching that smoking is bad for us and didn't understand why he was asking me to smoke with him. I teased him and replied," You're a bad example!" He and his family looked at me bewildered.

When I told Torg what had happened, he laughed and told me the Elder was asking me to pray with him! Since then I have smoked many a peace pipe in prayer. And the next time I met the Elder, I had a gift of tobacco for him!

What about Shaman mix cigarettes?
No, ceremony only. Would you bless a drinking fountain?

If cigarettes are evil, why use tobacco for sacrament if it is evil?
Both sides use things of power for sacrament.

Are you OK with my way of paying tithing?
You give more than your share. Offerings are essential to salvation. Buy a dress.

After reading the Mentina archives I wanted to pay tithing the way the Nemenhah did. They used the law of consecration where everything they had, even their time and talents, they considered to be God's. They would listen the the spirit to know how to use their assets.

I keep yearning to write songs and sing, is part of my mission singing and performing?
Your calling is to get the message out by any good means. You are doing very well. We are pleased. Do many blessings and clearings.

Any money coming?
Don't worry about money. Have fun!

It's hard not to worry when it is winter and the heat almost got shut off!
Is it shut off?

No, but you wait until the last minute!
Sorry. Why no faith?

So if I don't worry anymore, you'll just pay it?
Yes, only builds faith when emotions are attached.

What color is your hair now?
Yellow. I like change. Others here do too.

????

I'm not OK with men abusing women -

We 're not OK or wish it either. We cry too, and bring women home to Us. Please trust and wait for understanding.

Women can be unrighteous. too. Women are tested with unrighteous dominion over kids.

Do You think I have exercised unrighteous dominion?

At times, so did I. I repented and healed; so can you.

It seems we should be angry about abuse!

Be angry, it is God's will. It's the essence of Godhood to experience evil and grow out of it.

So we should just ignore evil and hope they grow out of it?

Of course not. You should fight evil. Honor people's agency, and when they do evil, stop them. Teach consequences and follow through. Your friend Shelly needs you to stand against her evil (alcoholism). Failing her mission is the evil. Alcohol is the means.

Good work with Steve. He needs a spiritual and rational anchor. We sent him to you for this. Like a teen, he needs boundaries.

What about the other two kids who showed up at Safe Haven?

Love them. They need you, too. You have more kids than you know. You are Mom and Dad to many.

Even Suzy?

Especially. She is a fallen angel. You help her reclaim her gift and return to her mission; connecting spirits with mortals.

I felt a lot of power at Safe Haven when I smoked the sacred pipe. I really did cast out a demon from my son. I hadn't done that before!

Yes, you have such a mother's hug. What could be more powerful? Kiss, cry, sing, pray, all works.

The message is love - Family is the essence of eternity.

The millennium is only the end of the mortal round.

58

Everyone from the beginning of time will have the opportunity for ceremonies of commitment to us.

Tell me more about Gays and lesbians?
A mixed bag. Every case is different. For some, caused by a Mom's unrighteous dominion. Others may be sick, chemical imbalance, infected mother while in the womb, agency. Some are a test and others it is their natural resonance. We are the judge.

Jesus' house is great. My people always build sacred first.

She was referring to a small building we were creating at Safe Haven with a white dragon sculpted on the outside symbolizing Jesus. It was inspired by the vision I had of him as a white dragon and also Safe Haven looks like a White dragon on google maps.

Thank you.
Needs flowers inside. 4 colors: red, yellow, green, white.

Chapter 7

Love sets boundaries

I was going through a tough time with a lot of people in my life. Between the people I rented rooms to in my home, people showing up at Safe Haven going against our values, and family members creating drama, I was ready to tear my hair out! One of our Safe Haven members suggested we read a book called "Boundaries". We decided to read the book as a group so we could support each other in setting boundaries.

I'm frustrated with one of my family members.

You need to set boundaries. I have that problem too.

I thought I was doing better .

You are. I'm still learning boundaries too.

Should we have a codependent support group?

If the learning isn't hard, it doesn't work as well. Learning the hard way is part of human school. I've got your back.

Why is it so important?

Relationships are the core of the universe. This is your family.

Love sets boundaries.

Was I OK with my kids?

You had good kids! Lucky, no need for curfews, room checks etc. You would have freaked out, now time for you to freak out.

I don't want this! I don't have time!

No you don't!

I'd rather spend my time writing about You.

This is about Me, relationships! I had to learn this too.

Maybe I could write about how You had to learn to set

boundaries?

It's the hardest part about being a GOD. When to spank, hug, or ignore. You made a boundary choice with Shelly with her alcohol problem at Safe Haven. It saved her life and helped the community. It was the major reason she got help.

I get angry when my boundaries are walked on.

Set before anger.

Is it hard to watch kids suffer consequences and not "Save" them. At what point do you jump in and save them?

When you pray, the consequence of prayer is less suffering. Repenting brings less suffering.

You say the family is the most important part of the millennium and Your kingdom. What is the best way to bring families together?

1- Teach kids at home. Creates bonding and respect for each other.

2- Don't lock seniors away. Take care of them in your homes. Seniors have wisdom and patience.

3- Give birth with no drugs. Bonds the family.

4- Mom and Dad should always date and make love often!

5- Do things together.

6 Shamanism is a family affair. Most important of all of them. Get the word out. Teach, "All God's People are Shamans".

What is your description of Shamanism?

Many types. Your type is an Earth Connected Grid Walker.

You want all people to be Shamans ...

Walk consciously. Some see, some reveal, some prophesy and some heal.

So like the description of a prophet? Would You say you want all to find their specific spiritual gift and talent?

YES!

What do we call an evil Shaman?

Sorcerer.

I had been asked by the elements to teach Shamanism. I guess this was another reason to teach.

I thought "at one ment" meant "righteousness".

It does. Would you rape the earth if you were being righteous? An evil Shaman is not possible.

The onyx from Giza is being ground to powder and sent to 28 nodes by Isis. (The Shaman from Giza)

We had been told by our guides a few years back that there was something special left at the Sphinx by Enoch for our time and that we would know later.

Onyx is a special crystal that enhances spiritual and Shamanic activities. We had been gifted with many small onyxes, which we then gift to our students to enhance their journeys.

This week we found out from our guides there was a Shaman working at Giza, called to find the Onyx blessed by Enoch and crush it up to send to as many nodes as possible to anchor the earth grid in female energy for the shift. They explained that he blessed it on behalf of Heavenly Mother for Her time. Female energy is protective and would protect the earth during the shift.

We had gone out of body and helped her locate the onyx. Unfortunately the dark side had their Shaman there, too. He was her boss at the Giza visitor's center! During that week the evil Shaman was distracted by the civil war happening at the time and wasn't shielding the onyx as much so it was a good time for her to find it.

Evil loves fighting and conflict, so he was excited by the fighting and liked fueling the flames. He was pretty upset when he lost the civil fight and the onyx too! He quit his job the next day at the visitor's center because he was so upset.

How will she do that?

However she wants. They will bind deep to Gaia at the nodes. Nexuses bind the surfaces to the net.

Feb.13, 2011.

Hi Mom, I've been seeing lots of Butterflies.

> *Symbol of butterfly means transformation. Stronger mission, I got your back. You are growing.*

I don't feel like I am doing much.

> *Doing lots. Look, you watch. You will be doing lots of blessings. Money soon, you passed the test.*

What test?

> *To not give up. Finish Shaman kit, do blessings and ceremonies.*

Feb. 22, 2011

Sorry we are late, we were helping my former mother-in-law move into her new apartment.

> *She's a good friend. She loves Us. She'll be easy once she is settled.*

I thought it interesting that She considered her to be a "good friend". My former mother-in-law plays the organ in her church and is very God centered. She also plays and sings amazing Southern gospel tunes I love to hear. I'm glad to know She takes care of Her "friends".

Is the atonement more about "At one ment" rather than some heavy need to save the world from sin ?

> *It's both. Yes, it means, "At one with Us." Any sin, no come home. Sin means not one with Us. Anytime you are not at one with Us you are in sin.*

> *Jesus was the first one to be at one with Us His whole life. For most it happens after death. For some earning the right to come home happens there. Do your calling to be Ryxi, leader of Goddesses with no anger or jealousy and you can come home.*

What is your definition of evil?

> *Sin is an action. Evil is a condition. A condition of not being at one with God.*

64

Did Torg's dad earned your kingdom while he was here before he died?

Yes, two weeks before. You have a high probability if you don't hold on to anger.

So does that mean I can't be angry or jealous about something like my husband sleeping with someone else?

No that's OK. It's not jealousy. Don't be jealous about others writing books about Us. Others are doing all that you do. Others having website, etc.

That seemed like a silly concern. I don't see myself being jealous about that. Unless they know something about me I don't know.

It just means We are all in tune.

Jesus' mission was essential. Suffered even though he hadn't done anything wrong. You sin and not suffer. He suffered with no sin. Now in balance. Absolute need for balance! Every tit has a tat. Every crime has a punishment. Every action has a reaction.

Some People have it wrong. Immortality and eternal life are not the same. Jesus gave immortality. Humans earn eternal life. (Becoming a God) Repentance is now possible.

I have friends who argue that even Satan can change and we should pray for him and keep the candle of hope lit for him.

Can Satan change and be righteous?

No, he chose to fight while in the light. (full knowledge). *Have to fight while in the light to be irrevocable. All his people too. They knew what they were doing. Same for you if you get light and fight. The reason you came here was to sin with a plan to reverse. To learn and be tested.*

Torg: "So sad to lose so many."

Other earths/worlds lost less. They also graduate fewer Gods. All less evolved realms, yuk. (other worlds).

That's how I feel about church sometimes!

Absolutely! You send out ripples from this church to change others. That's why I need you to go. Are you up to it? Simple callings don't make strong Gods!

Chapter 8

Agency

I went with my friend Earl to court to support him in his case to protect the constitutional right to free speech in public places, specifically art and music.

How do you feel about Earl's case?
I admire him. His calling is to preserve agency. His actions are pure. He is a son of God.

What do you think about the different political parties?
I try not to think about it.

Are you republican or democrat?
Neither.

What would be your favorite party?
Natural Party.

What do you think about the Globalists (Luciferians, New World Order) trying to make a one world religion and using earth based religion as the focus?
It is the "New Wave" and they want to control it. They want to control Shamans who take care of the earth and practice earth based spirituality. The age of Aquarius is My age. Evil is not stupid, they know it is coming. They don't want to stop the change, they just want Me to lose.

How would you lose?
By them controlling it. Not Me in charge.

How can they be in charge?
Remember the Crusades? They used the church to do evil things, like burning good witches. It happened slowly. Evil is patient.

How can we stop it?
Stay in charge. Watch Earl, he knows.

You mean keep the Constitution intact?

Part. All we need for evil to win is for good people to do nothing. Agency is essential for Gods to be.

How can they use Shamans?

Confuse them as to who they serve and how. The Crusaders started as good people.

How did it happen?

Very subtle. You need money. When does that become a want or a worship?

You tell me.

There is no line, it's individual. Becomes a craft.

What is the difference between "Hood" and "Craft"? (Priesthood vs priest craft and witch hood vs witchcraft.)

When it is done for gain, not love.

Other Shamans receive money for blessings. Usually donations. I'd rather do it for free. I worry about doing house blessings for money and falling into "craft".

You're heart is right so far. You have expenses, accept donations.

Some churches believe we shouldn't charge at all for blessings.

Yes, they want to be safe. Slippery slope.

What is an example of confusing Shamans?

How far they will go to protect Gaia. Wars to protect are evil. Block roads, spikes in trees, metal in machine oil, etc. All wars start small.

What would be the best way to protect the land?

Pray and ask Us first.

Was I too sexy last night at my birthday party? I was showing a little cleavage.

So do I. I liked your hair. Wow, last night! Joyful to Grid. (We danced. And when we dance the Grid goes up.)

How are we doing with my former mother-in-law? (She was struggling with family drama, generational healing, *etc.*)

It took a long time to get to this place, it will take a long time to get out.

Am I loving her enough?

Yes. Not for her, but enough for Me. You're doing too much for her now.

??????

Why did you allow the Native Americans and so many other indigenous people get clobbered by the whites?

A lot of natives are better off now, but with destroyed agency. The people had warped ideas of Indians.

Why are white people so evil?

Power. It was not always like that.

Do You ever care, or is it just an attitude of, oh well, people will be people.

Don't insult me, please. I care!

Where is the justice?

You look outside with a narrow view. There's always justice. All deeds all the time. Good or bad don't escape.

During the Dark Ages, whites were nothing. The Chinese, Nubians, Arabs, and such were worse than anything whites have ever done. Made Us cry.

We have to wait until evil is full before acting. You're a mom. At what point did you discipline your kids?

I try to start as soon as they start to do bad.

We, too. We sent prophets to warn, and the signs were obvious to the educated and spiritual. Those that hurt or die, the innocent, come to Us.

So we just happen to be in history where whites are on top?

It's changing rapidly. First time in history where no one's on top.

Did this just happen, or did You plan it?

Nothing "just happens". You know this, this is your calling. Agency doesn't change big sweeps!

Do You create the big sweeps, and then it's up to agency how people react?

69

*Close. We see the future and try not to change it.
Otherwise it's hard for you kids to learn. We try not to step in,
but sometimes We have to.*

Example?

*You and Torg's money now needs pushing. It's a
critical time, we can't intervene yet.*

**So You're not helping the people who are getting our
money?** (Torg's business deals)

Not yet! Stages are in process.

**Why wasn't I born in a family that had money like Jesus
was?** (His uncle)

Most need trials. Patience a little while still.

March 6, 2011

*Hi, good Safe Haven meeting. Shelley is still sober.
All the group is learning.*

We told Shelly she couldn't come to the Safe Haven meetings
drunk. She was trying. We were learning from our boundaries
classes.

When does fashion and jewelry become out of balance?

No need to adorn.

What about my necklace? (I was wearing a turquoise
necklace given to me by a Hopi water clan chief)

No, power tools are not adornment. (The necklace
was powerful because it was turquoise and had meaning because it
was given to me.)

What is your description of adornment?

Does the item have purpose?

What about having a purpose of being pretty?

*No, you were made pretty. Jewels are for ceremony
to show authority. Used as a badge and to protect.*

What about a pretty dress?

*Slippery slope. What is inside you is what you are,
not your dresses or jewels. It's good you question. You're not
lost yet. Can't bring it with you.*

70

Women won't like that one!
>*Big problem, yes. Many people there have a hard time with no stuff.*

Why do You color your hair so many different colors?
>*Variety.*

I planned to take two pair of shoes back, the rest I have used all week. Did I go overboard?
>*Yes, not much though. Ask yourself: is the world better, your calling better, family better?*

If I feel pretty wouldn't I radiate and make the people around me feel better?
>*If you need to justify, you've gone too far.*

Later She would say She went too far about adornment and that some is OK. I still wear pretty dresses and pretty jewelry in spite of Her council. I just make sure I don't go too far.

What is the purpose of the earth burning at Armageddon? Is it Human created, natural or essential?
>*It's essential. A ceremony to cleanse. You get the gift of the Holy Ghost and Gaia gets the gift of cleansing.*

The whole earth?
>*The surface.*

Will everyone be burned?
>*Of course not. The righteous and prepared won't die.*

How will they survive if the whole earth burns?
>*They will be protected by the Priesthood, water, or caves. They will be in certain locations where they will be safe.*

Don't we already have the Holy Ghost?
>*Not all.*

So those who don't have it will get it?
>*No, those who have earned it will get it. Some unrighteous will survive the burn.*

When?
> Can't say.

Are there sacred numbers?
> 1- unity in purpose. 3- To be stable and trinity. 7- creation. 12- Useful mathematics.

Why is 666 the mark of the beast?
> Can't say. You'll know what it is after it happens, as proof.

Many people are planning to starve rather than get the mark of the beast. Should people be willing to starve as opposed to getting the mark of the beast?
> Twits.

I've been supporting Earl all week in court on his case for free speech and he won! How do you feel about it?
> He won agency.

Did you help much?
> Didn't need help. You kids did well.

I thought you helped?
> We don't want to help. Agency makes you a God. We help when you need it. He did very well, he needs to do more.

Anything else you want to say to him?
> Love you. Do more art.

Our guides said they couldn't follow us into the court house because there is a shield over it. What is the shield?
> Bad shield, yes.

Can you get in?
> Yes. Your guards can't.

Can we take it down?
> You're not strong enough yet.

Why don't you take it down?
> If I fix all things, Satan I would be.

Why is it OK for us to fix it and not OK for You?

It's the way the plan works. I'm already a God, you need to learn how to be a God. If I carry my kids, they never learn to walk.

Who is the Anti-Christ?

The Anti Christ will be obvious soon. Can't say who and when. Yes, we know who it is.

How do we get rid of evil politicians?

Vote them out! Good ones need allies.

I thought you said none were good?

All established parties are not good.

Was the Japan Earth Quake caused by Haarp? (Man-made technology that controls weather)

No.

My guides say there are other sacred earth sites for gathering people like Sanpete County has been and will be. They say Sinai, where the Ark landed, is one of them. What do they have in common?

They have onyx homestones (stones to guide home and anchor).

Why?

They are human sites, not Gaia or grid sites. Places to bring humans together.

Where are some of the others?

Mecca, a secret place in Tibet, lower end of Lake Baikal, Australia north of Sydney and more. 29 all together. I chose and blessed them for this purpose.

Why are there only home stones at Adam's temple?

The whole valley is blessed.

Home stones are stones, usually onyx stones left at sacred sites by people visiting and wanting to have a spiritual anchor there.

Why not Eve's temple?

Couldn't get there very easily. A good place to teach mysteries.

My friend Greg wanted to test if I was really talking to Heavenly Mother so wanted me to ask this question....

What would be a good question for Greg to ask?
If I exist.

Later when I told him Her answer he seemed surprised, because that was the answer to the question he had asked in his mind. I guess we passed the test.

March 20, 2011

How are you?

Watching kids. I like Rider.

She's your daughter.
In whom I am well pleased. I am pleased with you too. Sad about Karra Dropping her calling.

Our friend Karra became afraid of what we were all doing; talking to angels, Heavenly Mother and other worldly beings and decided to quit our group. Even to the point of no contact at all with any of us. We were sad. I don't blame her, we are really out of the box! It can be scary to break our paradigms.

Torg's dream?

Torg explained he saw himself with a bunch of our friends before we were born getting our calling for our earth walk. We looked like liquid crystal. He recognized us by our essence. Karra had been one of them.

Any comments?
Nice Kid, don't mess up.

Is it true there are 200 of us?
A few more. Less than 1,000. Satan really wants you. Select group saved for the change over to the New Millennium.

Did we have these natural abilities as intelligences or did we develop them later?
Both. Torg was first a dragon. Finn wouldn't take that chance.

74

Finn is our dragon friend who is a healer. He shows up when we need healing.

So we are chance takers?

Yes, Gods. We asked Prescott to stay untill the end. He will be a God.

Ryxi, you had a question. Cigarettes are made for fun, not sacrament.

I didn't remember asking that. I remembered that in the sweat lodge over the weekend I wondered about it, because the lodge leader went out to smoke a cigarette and I remembered other Shamans doing it. I wondered if a cigarette would be considered "sacrament" like smoking a pipe is. She must have read my mind.

Why did people live longer in Adam's time?

Was in the genes. Needed at the time. You can now if you turn the genes back. Not a good idea. It's not time for long lives.

I took a women's self defense class. The teachers told us we should maim or kill someone who tries to attack us. What do you think? Is that Christian?

Make them wish they were dead. Yes, push for women defending themselves!

Which subjects for the book should I cover first?

Sex, O wow! Making spirits. Fun in Heaven! Don't want to be the centerfold! Ha ha!

The void, Big Bang, lot's of smaller bangs (earths etc.). Intelligences.

Can people handle your humor?

Not funny, I was serious!

What is the void?

Infinite supply of intelligences. All Gods make holes and create universes.

75

Where are resurrected beings that aren't on Your planet?
Other earths, space, Sundance, anywhere.

How was the earth physically created?
Matter is intelligences. Intelligences acting on law. Atoms and intelligences. Groups become water, etc.

Why is singing so important?
Music affects all, not just life. Resonant harmonics affect the crystal structure of the universe.

Are there other Celestial realms right now besides yours?
Not here in this universe.

What were the false starts?
Star on forehead.

What is your definition of "Raping the earth"?
Take more than give: rape.

For the Millennium, how will you help Jesus?
Jesus is the God of this world here on GAIA.

Will he be here physically?
Except while on vacation. I'll be telling Him what to do.

Where does He go on vacation?
Surfing on other planets.

Will Your time be about healing the earth too?
She can rest for a time because you Shamans will take care of her. She won't have to fight pollution, etc. That's why I need everyone to be a Shaman.

What are the biggest challenges families have today?
Competition for time. Busy lives. Parents' lack of commitment to each other. Teach in the home.

Why wear white in heaven?
Pure.

Was Lot's wife really turned into a pillar of salt?
No, turned into a pile of ash. It was very hot because of the tar burning. Really hot, hot volcano. We ordered it.

Angels did it. Shem (Melchizedek, translated) asked the dragon to burn the cities.

What are the new skills, traits and needs for the Millennium?

Shaman skills, family skills, restore family.

How would Satan take over your time?

After he is bound, he can't. One bit at a time. Make warriors weak by playing on emotions. Empathy being misunderstood. Help too much. Take power.

Co-dependency?

Mothering, smothering. Star for Ryxi. Even on your own without Satan's help many of you have trouble with this.

How will Satan be bound?

Priesthood. Both women and men blessing and working together in harmony. One alone hasn't worked. Men alone and women alone has been tried, doesn't work. A process that's starting now. Has been going on for years. Yes, you help. Yes, Yes. Started 500 years ago.

How?

You people started to ask God, Martin Luther, Calvin etc.

Were women using their priesthood?

Not enough. After the witch hunts, people started getting a conscience. Ann Hutchinson helped.

I'm getting really tired. I have a lot more questions since we haven't talked for a while, but I'm too tired to ask anymore.

I'll come back tomorrow afternoon. Sleep.

Love you.

Chapter 9

Resonant Harmonics

March 21

Thanks for coming again.

You're welcome. I tried chocolate, love it.

Why is singing so important, again?

Music affects all, not just life. Resonant harmonics affects the crystal structure of the universe. Much of Genesis was caused by music. We sang and played to assist Gaia to exist. The sung word is better than the spoken word. You were there. You sang when Jesus was born. We weren't just singing to be happy. We were performing a ceremony to anchor that space and time.

Were we all singers?

Not everyone. Spirit resonance washed through harmonics both physical and etheric.

Translated and resurrected beings sang with them? (spirit singers) Chorus?

Star on forehead.

I sang well then?

Did, might again.

You don't like my blues voice?

Teasing. You should sing to all elements and directions.

Do elements sing?

Yes, Ryxi , you felt them singing to each other in Moab. Listen, you will hear!

I had been with my family 4 wheeling in Moab, Utah. We had stopped and I felt this interesting energy coming from each side of the canyon connecting through me. I thought they were both sending some kind of energy through me. Too bad I couldn't hear it.

SIng to house, car, tools, food, tap water and to be healthy. Distance singing is less affective. You will soon sing at nodes, nexus' and sacred spots. You sang at Green Mesa.

When Torg turned the rock over at the node in Mesa Verde, he couldn't let go until I sang and dedicated it to Jesus Christ.

Torg twit, next time sing first! Needed both, singing and dedication to let go of the rock. Dancing is important but not as important as singing. Sing major keys, better than minor. Happy, not sad. Just sing! (she started dancing the infinity sign. Funny.)

When we sang to Lake Baikal did more algae (life) grow or did the existing ones come alive?

Both, one of your top works. High Five!!

A few years back, there was a time when we were asked to do ceremony and sing for differently nodes. They needed extra energy to become stronger. We were told that Lake Baikal in Russia was dying and if we sang to it the algae would have the energy to grow again in spite of all the poisons being dumped into the lake. As we sang, we saw in our minds eye, sparkles falling from the sky into the lake, descending to the algae and the algae starting to come alive! We kept checking in later and through time the spirit of the lake grew so big it over flowed into the surrounding forest. We were told other Shamans were there assisting and that soon the paper manufacturing company's that were spilling poisons in the lake felt uncomfortable and were moving from the lake. Mission accomplished!

ALL my favorite music artists smoke pot, like Bob Marley, Sole etc...Why is that?

Looking for answers. Mindset. It still makes them stupid.

Does it open creativity?

Yes. Talk to Me would work better.

Is singing female?

Yes, but all intelligences sing. Female energy is singing and dancing. Male energy is knowledge. Both do both. Men sang at Eves temple. Women studied at Adam's temple. Prescott (orbiting earth) likes to dance but can't sing.

Who was I in a past life, did I write?

Yes, you want a clue? It wasn't a living language that you wrote in the past. Olmec, close to Adamic language.

Which continent?

South America. Olmec writer. Wrote on papyrus. You were an emissary and traveled all over from Mexico to South America.

I was female?

Very

Was I pretty?

To the Olmec. Looked African.

Husband and kids?

Yes, no more info.

Did I have other lives other than this one?

No comment.

March 23, 2011?

Hi, Torg needs me to comfort him. Bad night the other night. Needed to be alone to pass test. Jesus passed the same test. His mission overwhelms. You'll need all your inner strength soon.

I had noticed Torg listening to music a lot lately and crying. He did that a lot, but he was doing it more than usual. He finally confided in me that he had thought of committing suicide the night before. I felt bad that I hadn't picked up on it. He said he felt overwhelmed by his mission.

Ryxi you did right by not comforting Torg. He needed to be alone. Love you.

Torg, I hugged you through music. You are ready for the big fight. Both of you at unification ceremony. Both of you exercise, lose weight, get well.

Write stuff for the community, build soon. Watch dancing with the stars, less crime shows. Magic shows are drawn to you cause you are magic. Bad plots. Get rest and get well. I'll be back Monday, you need Me.

We do!

I know. I AM, bye.

April 10, 2011

We watched a show on the history channel about Hiawatha of the Iroquois nation meeting a man born of a virgin who healed him and he went around to the other tribes and the nation was united and healed. Who was the Iroquois peacemaker? He gave them the great law of peace.

One of Enoch's people. A translated being. Not born of a virgin. It was obvious he wasn't mortal. Didn't ever eat anything. Had no fear. Walked on water. Surrounded by white light.

Who was White Buffalo Calf Woman?

Another one of Enoch's people. A translated being. It was her mission. It happens a lot. Enoch's people are helping. You've met some. Trust the stranger with a message, he may be an angel.

Example?

Los Angeles. He told you to go home.

I don't remember. Before I got married?

Yes.

Was he supposed to tell me to go home?

It was his idea. He didn't know you were supposed to marry your ex. I told him no. Sorry, rough ride. You did well, but didn't listen to angel. Never not listen again. Ask Us.

What did he look like?
Dark hair, brown eyes and tie die shirt.

I vaguely remembered someone in one of my acting classes telling me I should leave LA, that it chews people up and spits them out...maybe that was him?

Any others I might remember?
At festivals. They give hints. Torg saw one who gave him a blessing.

Torg: I never saw him before or since. He came to a party and said I needed a blessing. Said I had to suffer more than Job.

Why am I attracting all these people with addictions and problems?
Humble service. Training.

What am I supposed to learn?
When to serve and how. Remember to ask God.

Haven't I been doing that?
Mostly.
When didn't I?
Letting Margret move in wasn't the best choice. There is hope for her. She's leaving. Your ripples are good for her. You are a Good Samaritan. Yes, you serve the least of these. Better renters are coming. A team member is coming.

I needed the money, that's why I let her move in, besides wanting to help her.
Bye, Love you.

Chapter 10
The circle of life

April 11, 2011

Why is Nature sometimes so evil? I saw a program about nature where a baby Hippo was attacked by a male hippo and was running behind his mom who was trying to protect it. The male finally got hold of it and savagely crushed it in his jaws. It was horrific!

Let nature run. It was made that way. The Hippo killed the baby so he could have a baby.

You can evolve a rose, why not a good hippo that doesn't go around crushing babies so it can have sex!

Roses kill other plants to live. It's an old model that will soon change. All plants and all animals become the way they are through evolution.

Why have that model in the first place?

It works best. All life exists in competition.

It seems there would be a better way than competition. This seems evil!

How is food made? You have to eat food life to live. They have to compete with weeds.

I was so upset with the conversation and was getting angry I took a break so I wouldn't lose my temper!

How will the Lamb lie down with the Lion in the Millennium?

It's metaphoric. If that were the case there would be no more Lions. There would be no Lions, cats, Dolphins, Whales because they all eat meat. There would be no animals at all. All eat other life.

Humans?

Like I said, all metaphoric. There would be more harmony. If a lion changes and eats straw like an ox, it's not a

Lion. The teeth, hair, muscles and stomach would change because it is used to eating meat.

Why not us, we eat meat?

You are wise enough to make changes and eat just fruits. There are some fruit eating animals. It is best to eat plants, mainly fruits.

Will most humans die?

Most will not if became fruitarians. People with bad diets are more likely to die. It's relative. A good diet keeps you healthy.

So healthy people will survive the change into the new Millennium?

Yes, meat is OK.

How do you feel about the New age and Wiccan philosophies about the Age of the Goddess?

I like some of it. I like the gentler life and the Shamanism. I don't like the drugs and that they deny Jesus. It's a big movement. There are many kinds. Balance would be best.

What did you think of the Goddess pageant I went to last night?

Silly. It takes more than a white dress to be a Goddess.

April 12, 2011

I was really upset about the conversation about the hippo yesterday and actually cried myself to sleep the night before thinking about how unfair it is that life has to do such horrible things to survive. I was hoping to hear that this was not supposed to be how it is. That maybe Satan had influenced the animal kingdom somehow. I actually wasn't sure if I wanted to ever talk to Her again, but I finally did.

Do you have something to say?

Yes, plants and animals. You make no moral judgments. You don't understand. This is how life works in all universes.

It's sad. It's a dog eat dog universe!

86

*All spirits come home. All life force goes on forever.
No true death. Only a mortal view.*

Seems traumatic.

*Life is. Life would not be without it. Opposition drives
everything. Plants serve animals. Animals serve others.
Plants serve each other. The sun powers the circle of life.*

*Some people try to keep bodies from returning to the
circle, not righteous.*

No embalming?

No, the body is only a place for the spirit to live.

What about Cremation?

*Better than embalming. It turns to minerals. I'd prefer
you didn't. The last gift of life should be to the circle of life.*

Did you inspire the movie, Lion King?

Other Elohim did. Feel better?

**Kind of. I still feel bad about and adult male hippo
crushing a defenseless, helpless baby in it's mouth.**

*Me too. This will soon be solved. No fear, Dolphins
will not go away. All animals will stay. Some come and some
go always. We'll lose some.*

Can you give an example of ones that won't stay?

*Shouldn't. Anything with thorns. Plants have thorns so
animals won't eat them. Only eat enough so both can live.*

**Can we evolve a rose so it doesn't have thorns and we
can keep them?**

You people help? It will be fun to watch.

**Will this be the first time this will happen in your
universe?**

Yes.

**Will other earths in this universe go through the same
process as us?**

Yes.

Will some stay?

No, all evolves.

What is your definition of glory?

Highly evolved places of light.

April 17, 2011
Can Satan influence animals?

Swans and penguins don't act like that. Swans do kill babies for less competition for food. An evolved example would be a beaver who mates for life and helps his kids make homes for themselves. Satan uses stress over resources to control animals. Hippos usually don't kill, only because of stress over resources.

What about the concept of abundance for all?

Abundance not yet. People aren't advanced enough yet.

So it's OK for Humans to kill other humans for the sake of resources?

Of course not. But I would consider it as a factor when judging. Lack of resources make animals fight, kill babies and hoard.

How would they act if Satan didn't influence them and they acted on their own?

Some would fight, some not. Some would hoard, some would share. Humans too. He pulls, we pull. Opposition in all things.

Are there certain animals that are easier to influence?

Some snakes are easier to influence. Meat eating animals are easier to influence. Solitary animals are easier than herds. Burrowers are easier than animals who live in trees. Water dwellers are easier to influence than earth dwellers.

Can Satan influence an animal to attack a human?

Sure.

It is easier for Satan to influence water.

How?

It's a matter of physics. Info spreads easier. Molecules are not close together in the air. He uses air and the winds too, but it's harder. The Santa Anna winds are an example.

What is so special about crystal?

The structure. The way the pieces fit together makes feedbacks happen. Do you understand resonance? Microwaves, water, and oils.

My friend Paul saw me in a crystal house during a healing session. (Matrix)

Wow, can make you stronger!

Our planet is a crystal. Yes, many kinds. Much of life is made of crystal. Bone, blood, eyes, protein, DNA.

April 24 , 2011

How was your Easter?

My favorite was the Jesus Rock Church.

Figures, you are my daughter.

How was yours?

Great. No Easter eggs. Lots of chocolate. I don't need to eat but choose to. Praise God. High five to Jesus.

Did he come back from his vacation for Easter?

Yes, good kid. We danced and sang.

I danced with Jesus in a vision.

You sang for him too.

In a choir?

Individual and in a choir.

What do you think of the Rock church I went to. It was soo fun and I felt the spirit.

Yes, it's fun but lacks full truth.

Don't they all?

April 25, 2011

Rider had gone to California with the intent to swim in the ocean and connect with the whales. When she tuned in they were mad and wanted to know why Torg and I weren't with her. She was very upset because she was excited to meet with them.

P's here:

Why are the whales mad?

Whales are mad you two are behind schedule. It is our wisdom you haven't shown up yet.

I thought we were ahead of schedule?

For the grid, not for the whales.

What are we supposed to be doing?

Joining is not just human. The guard at Safe Haven is talking to elephant Shamans. You should be too.

We are supposed to be talking to the elephants?

Yes, to all life.

How are we supposed to know this?

The Shaman animals from the Madagascar node take over. You are supposed to talk all the time. We thought you knew. It is a process. All three of you.

Why were they mad?

All three of you were supposed to come.

We'll have to apologize to the whales' higher selves.

It wasn't nice for them to be mad. They are fine now.

Who told them?

God.

Why didn't God tell me?

All life together doing nexus work.

Anything else for today?

Happy Gaia day! (Earthday) Renewal.

Let's talk tomorrow afternoon. Go on a journey. Remember, the whales are fine, God talked to them. No blocks.

April 29, 2011

The next time the P's showed up we wanted to get confirmation on where we had gone on our journeys.

P's here:

Where did we all go on our trips?

90

Renee went with Prescott, then squid, then whales, then to the humpback in Hawaii. Didn't find the squid.

Rider, found a Narwhal whale in the North Atlantic.

Torg went to see the elephant and parrot in Africa.

The turtle is sleeping now.

How do we connect with micro-organisms? (small life)

Science will give understanding.

Any we missed?

Yes: snakes, fish, spineless animals.

The elephant eats crystals to help connect.

April 26, 2011

How do you feel about mixing races?

I like it, but makes marriage harder.

Why do you like it?

It gets rid of bigotry. More growth is forced. Different ways make marriage harder. Different religions, philosophies, education, money, etc.

But doesn't that promote classicism and elitism?

Keeps people down, but helps keep marriages together. Both sides of the coin have hazards.

Any scientific reasons?

Mutts are stronger. Kids from mixed marriages are healthier but more likely to have single parents. Less healthy emotionally but more healthy physically.

Is the race mix why I was supposed to marry my ex?

Some. Lots of reasons. It's complicated. Single Mom, race book, etc. You would not be who you are and neither would they.

If you were married to a rich producer, how would your life be different? All part of your path.

So I had to go through all this so I could write a book about you?

Me and other callings.

Book about races?

All callings. Think big...Shaman, recorder, general, community, Mom. Think about how different your kids would be if they were raised by someone else?

How can people prepare for the new Millennium and the shift?

Have food storage, be off the grid, Study survival skills, live organic and raw. Shamans will survive and assist others to survive.

You said the Earth will rest and heal during the Millennium because the Shamans will be protecting her. Why else do we need to be Shamans?

There is the spiritual part and the healing aspects of Shamanism besides healing the earth. People need guides. It won't be a big magic bang. It will be a process. People won't be any different. They will still need teachers, food, clothing, etc. There will be more Shamans guiding.

Your Safe Haven Problems are a chance to learn. Those who are creating problems will leave if you stop allowing. If you stop reacting to them they will leave. Ignore them. Don't respond, It works. Only respond to good stuff they do.

April???

When Satan fought in the light, was he sent to prison automatically, scientific or just happens like when we resurrect. Or was he judged and God punished him and put him there?

All of you judged him. No possible other way. He had to go. Consensus. We expected it. Always happens. Needs to be.

He knew he was the one you expected to be evil?

Yes, his agency, his choice. You read Milton. Says it well.

Milton?

You find it. (in Paradise Lost.)

If you know it will happen is it easier to accept?

You haven't lost any of your kids. You can't feel it.

Yes, I can only think logically about it.

Pain.

May 4, 2011

Ryxi , you have more to work on. No one's perfect. Even me. I am very God!

Teach my children and Shamans. Teach truth.

May 8, 2011

Happy Our day! (Mother's day.) *Did you like the statue? I picked it out.*

Yes, Thanks. Where is it from?

Middle East, means "Bringer of Life."

The statue she is referring to is one Torg found at an Earthday festival where we had a Shamanism 101 booth. Torg felt Her guiding him to buy it.

We decided to take Mom and the elements seriously and start teaching Shamanism 101. I like to be efficient and thought it best to teach a group at a time. More fun too! We figured an Earth day festival (Earthjam) would attract the type of people who would be open to it. We had many great people show up at our booth. Mom said She guided the right people to us. Even planned our booth to be in a perfect location to be seen. We were excited about all the amazing people who showed up!

This first class of Shamans were very powerful. We bonded with each other and made heart connections that would last forever. We work with many still to this day.

Chapter 11

Maxine

May 9, 2011

Maxine- (My Godmother from the 1500's.)
> *Pomegranate has vitamin C, antibiotic.*

Why sacred?
> *Symbol of the blood of Christ.*

How are you?
> *Great!*

How is your boyfriend?
> *Great! My review for resurrection is soon.*

I thought it happened automatically?
> *Yes, the review is a progress report.*

Who do you meet with?
> *Two angels.*

Who are the angels?
> *The names are not important. It's like an academic review committee. Oregon grape root tea is good for the flu.*

Will all people resurrect? Won't some choose the dark?
> *All people will resurrect, even those who choose the dark; will resurrect for a while only. Only naked intelligences go back to the void.*

Can they go through a Big Bang again and start over in another universe with another God?
> *Not like here, but maybe. Hard to explain. Not another universe like this. A person choosing dark will resurrect temporarily, then lose their body and go back as a naked intelligence.*

But don't you have to heal to resurrect?
> *Only if you are celestial. Others complete at death, resurrect and go to the realm they evolved to.*

What about cats?
> *Cats are celestial. Only celestial beings resurrect now.*

At some point a mass resurrection with everyone?

Close. End of my Millennium, less evolved beings will resurrect. Many after Armageddon. Those going back to the void are gradual after everyone else. When they go back to the void, body and spirit stuff stay here.

Mom-

Walk out talk in trees tomorrow. You're doing lots at once, need to guard. Helichrysum drives out, sage maintains. Sweet grass invites good.

Pongust: - Please put Goddess statue on altar. Giza onyx and herbs in a bowl. Herbs, raspberry leaf, Shepherd's purse, ginger.

Why?

Female Power. Mom will come into a dream with you tonight. Deva in trees tomorrow.

May 12, 2011

Who's here?

M,N,F,P's,Isis (I am) G,S,H.

Isis?

You guys rock! Saw Torg, last visit. Came, saw, conquered. (1st Shaman Class) Visited Torg last night.

Torg said this morning someone came to see him. It takes work; going on journeys. He was tired the next day.

Were you there last night?

Yes, I opened the shield. Lots of guides were there. Ripples. No, more like waves. New recruit ripples. The dark should give up!
Dandelion, wow.

She's talking about the kind of dandelion where you blow it and pieces fly all over. I had picked one on my walk earlier.

Send out the ripples like a dandelion. Mom and you are a great team.

You did very well, you get a ten minute break!

All Consciousness is one. God and we are with you as one. Great minds think together. We are so good together.

May 15, 2011

Wedding fun, rest, rest, rest more. (We went to my Nephew's wedding)

Favorite part?

> *Wedding & dance. I like simple weddings, not commercial. Spend lots of money. Wrong focus.*

Torg felt a female spirit flying along with us outside the window of our car as we drove home. Who was it?

> *Alice (Torg's Grandmother) flew with you. You were sleepy, needed help. She moves fast! Fast from Our planet to help you and came back.*

Is she resurrected?

> *Yes.*

Why couldn't he see her?

> *She didn't want you to. Followed you from the front door as soon as you left.*

What happened to Jesus' body when He resurrected since it is apparent that we grow new bodies when we resurrect and don't need the old one?

> *Angels took His body back to Gaia. It dissolved. We did not allow his body to be corrupted, no worms. Dissolve is bad word, doesn't describe it. Then took the essence back from the earth to make his resurrected body. Used His earth body, a special case. Took His human body essence and made a resurrected body as a testimony. A person resurrects wherever they are.*

Where was he when he resurrected?

> *With Us.*

Your planet?

> *Close. You read about it, didn't touch, Jesus hadn't ascended yet. We met Him where He was.*

97

Hadn't met Father yet?

Was in spirit then. Later showed self to apostles, was resurrected.

How long does it take?

Started the process in two days and ended in three. Others on the God path take longer to resurrect.

When it starts, wham! He taught spirits for those two days. No more cake. Make you sick!

Hadn't gone to the family yet for 3 days?

Didn't come to Our planet yet, We met Him. Went to spirit prison, set mission. He wanted to hug Mary, couldn't. Two days to wait was an eternity!

He hugged her after the two days, obviously.

A lot!!! Took her places.

How did he do that?

He's a God!

I wanna be in love!

You get ready. I find someone.

What do I need to do to get ready?

Paint your fingernails.

I have red orange hair. Green dress.

I thought they won't let you wear green.

I AM, try!

I guess if you want to wear a green dress you should be able to wear a green dress! Try and stop you?

Damn right!

Are you changing the rules?

No, guidelines when talking to kids. I don't like purple.

Any thoughts for this week?

Have fun! That's an order! Eat, drink and be married.

May 23, 2011

To the elements:

Thanks everyone! (Helped me with a person issue)

You the Ryxi! You help the world not end...less floods, tornadoes etc. We don't understand Humans!!!

Why do you help some and not others?

Mostly we choose our assignments.

Do you get specific calls?

Mission calls are for the few!

What things do you do for GOD?

Guard, guide and send dreams. Throw stuff.

Chapter 12
The Guards

Mesa Verde in Colorado is a sacred site called a "node", one of the places that hold the net of Ley lines together across the planet. We took our Shamanism 101 students there for the summer solstice to drop off the onyx from Giza. We also wanted to do ceremony to light up the Grid for the summer solstice and to energize the Grid for the shift coming, Dec. 21, 2012.

We had gone there last year and opened the node and dedicated it to Christ.

While we were doing our ceremony Torg suddenly went out of body without warning and stayed for a while, causing us more than a little scare. When he came back he shared with us that he met other dimensional beings from another universe hovering over Mesa Verde, as guards for protection. They were upset God just left them there and hadn't communicated since.

The experience was so wonderful; he felt like he was a Dragon again dancing for the Grid. His body was pain free and his feet didn't hurt anymore. He almost chose to stay. The love of his friend, Rider, who was there teaching with us helped him have the strength and direction to come back to us.

June 15,2011
Hi kids, love the job well done! You knew that.

You're a good example of telling Your team "good job". I need to do that more often.

I need to say it too!

Tell us about the guardians at Mesa Verde?

They hadn't heard from Us since they took their role. They hadn't asked (prayed to Them). *Now we are checking in on others. Maybe a policy change is due. Agency on both sides. If we said "hi" more often Satan would find it harder to win. Hard call.*

You don't know the balance between checking in on them and waiting for them to ask?

The guards aren't our kids.

Who's kids are they?

> *Friends. Hard to explain. Different reality. You don't have the concept. Sometimes We help Them too.*

How?

> *With what they need. Masau wants to meet you.*

It all made sense now. One time while visiting Hopiland, Torg suddenly had an episode where he was clearly not in his body. At the time, we did not understand what had happened. Now we know that Masau was trying to bring him into his realm to meet him!

Masau is the guard over the Hopi area. The Hopi are very much aware of him and do ceremonies where he shows up. They have legends about him that can be researched.

> *You should meet one from each continent at least. Many more, not just at main nodes. Your Safe Haven guard is a good friend to Masau. All work together. When you graduate, you may also serve other realities.*

Did you?

> *Yes, two times.*

We can't see them.

> *You can. Would scare you too much to see. (Masau) The Hopi can, part of their culture. Balls of fire helped with Moses. Spinning winds helped Buddha. Quaking water helped Palau.*

Who is Palau?

> *A Malaysian priestess using volcanic mineral water for healing. It only worked when shaking.*

So what does Masau look like?

> *Best you not see him. Your guard much easier but your mind won't let you see. Torg can see because of his science fiction history. You're training others. Machu Picchu rock....Go to Hopi, meet Masau. Machu Picchu too.*

> **Before Pele?**

> *Yes, Hawaii is OK, mostly fun. Australia, Lake Baikal, Missouri before shift.*

Sorry to keep bringing this up, but are we going to get enough money before the shift to go to all these places?

Have to.
Why is it so important?
To meet the guards. Ryxi meet human Shamans. Kahuna, Isis.
Isis is a physical person?
Yes, supposed to meet her. She's nice, likes chocolate cookies.
I would like to see Masau.
You can look, but remember I told you so. Keep control.

This reminded me of one of my trips to Hopiland when I was by myself in my RV. My Hopi friends said they were doing a ceremony that night and were worried about me staying in my RV alone. I couldn't imagine why there would be a problem. They said they didn't want me to be so scared that I wouldn't come back. I'm not afraid of ghosts. In fact I would love to see one if I had that talent, so I told them I was fine.

They told me to make sure when I heard bells at 3:00 A.M. not to look out my window. Sure enough, at 3:00 A.M. I heard bells. I was tempted to look out the window but decided to take their word for it. Suddenly I felt the RV sink as if something had landed on the roof making the shocks lower the RV. Then it suddenly lifted. Freaky!

What do you think of our students?
No accident. This group was carefully planned. Three were called to be there but used there agency, freedom to choose, and chose not to come.
Kids did great. You taught them well, Obiwan. Should re-name Ryxi's sink. Energy shift when Ryxi go by because of Female leading.

There is an area at Mesa Verde where there is an energy "sink", a place where energy goes into the earth and comes out somewhere else, cleaned. It was part of a temple site where people would go for ceremony. As part of the ceremony, they would go to the sink and let go of things that didn't serve them anymore. The energy went into the earth where it was cleansed, then released at a "fountain" where the energy escapes from the earth. Shamans went to the fountain to bring more power to their prayers at this site.

There is a wall full of petroglyphs at Sand Island, UT, where the connected fountain is. We understand there are similar sinks and fountains at most sacred sites around the world.

We had a ceremony at the sink and buried the onyx from Giza and a rock from Lake Baikal. I led the ceremony as we opened the sink. Because I was leading, I was shifting the energy of the sink which also effects the Grid around the world to female leadership.

One of our students still seems afraid of me.

So am I. I tried disciplining you, no results. You not do well, get mad and get even.

I'm good now, aren't I?

No comment.

Your students are your higher family.

How many blonde Shamans does it take to raise a family? Two, if one is me.

There will be a Dragon dance....where?

Can't say.

Torg's lifetime?

Yes, to reset Grid. The whole Earth will not burn.

Good to know.

Thought you did. Volcanoes, global warming, bombs, not the whole. This is now a ceremonial burning. The burning of the Earth starts the Millennium. This is the time of heat. Rock and roast. People will be fine as long as they follow the spirit. Most people will be fine. More damage by water.

Do you mean there was more damage by the Flood or we will have more damage now?

Both. Side effects of both.

My former mother-in-law is struggling financially and needs to move away. I offered to help, but she still wanted to move. I wish I had more time to help her. I could have done better.

So could we, with creation. But did enough. You did enough, keep doing. She will need more blessings.

Should I get books on dating and understanding men?

I never figured it out.

Are you finding me a Shaman to fall in love with?

You're not Shaman ready. Find a fun guy though.
Masau likes you...

Wow, teach mysteries.
Love you, bye.

June 16, 2011

Me, Gods and angels here. Grid is still well.
You need to become a scholar.

Subjects like the Kabbala or other mystical writings?

Some mystical, some pragmatic.

Chapter 13

Be prepared

June 25, 2011

I was given a book to read written by a woman who had a series of dreams about the future called *"Through the Window of Life"*. In the book, people who listened to the spirit would know where to go to escape various situations like invading armies and natural disasters. They would hear the spirit tell them to set up camp in a certain place and then the next day the spirit would tell them to move just in time to miss an army invasion. The author saw armies from the Tribes of Israel protecting the people of God with high tech weapons as groups would travel to safe places in the mountains.

There were times when the rains flooded the land so badly there were rivers flowing and kids would be caught in the currents. People would pray for help and see angels carrying their kids to safety and then disappear.

The main message of the book was to listen to the spirit, be prepared with our physical needs and share. When the people shared with others in need, their supplies would keep refilling. When they were greedy or afraid, the supplies would get bugs.

What do you think of the book, *"Through the Window of Life"*?

It's true but broad. It's about the big picture.

Is that how it will generally happen?

Mostly.

What part isn't how it will happen?

It was her interpretation. No specific time or place.

Do you mean it was a filter through her eyes?

What she relates to. She saw things from lots of different possible futures. Lots of different people. She stood proxy. Packed lots into one episode. Many scenarios will happen soon, not all to one person. There are other cities of light all over. The tribe of Issachar will help. Also many others. Some are in one place.

107

Can I have more details like when and where?

Won't say. There will be different armies from the different tribes helping. People will help themselves. The sequence of events is correct in broad strokes. Foreign troops from here. They are speaking a military language.

Also, there will be New World Order troops. Rothschild, Chinese, Africa, hired troops and mercenary troops.

Will Safe Haven be a Safe Haven?

The mountains, yes.

Safe Haven is near the mountains, will it be safe enough?

Why are you worried? Do you pray? No worries then, mate!

Are caves in the area a good place to hide?

No, you know how to hide in plain sight?

One of the armies sent to protect the people of God in the book was from the tribe of Issachar. Do my sons have Issachar blood?

Yes, lots of tribal people are not missing. Pieces of all tribes of Israel were in Judah when the North was conquered.

Are there other tribal units out there getting ready to help?

Some. Naphtali, Gad, Asher, enough.

Do the armies have all the advanced machines like in the book?

Some. Still a lot coming together, will be ready to help. They can tell if the heart is pure.

How?

You can, you should. You should pay more attention, you are still doing well. You give freely, don't hoard. You need both. You need to train to fight, also to give.

Go with spirit?

Oh, so true, Obiwan!

Angels are waiting. All over like Moses. Look to the snake on the staff. (Jesus) Can't store Manna, except on the Sabbath.

We could use manna in our cans...

You've got soup.

My patience is really being tried right now. I'm really having a hard time controlling my anger with one of my renters who keeps ignoring boundaries.

Talk to him without anger. Never get angry unless you have the wisdom of God, good luck though!

I have blue nails. This mom likes Reggae! Brown yellow, pink striped braids.

Swim in the Salt Lake. Go to spiral jetty, it's magic!

Your student, James is ready for more. Ryxi , you are unconsciously sending him your face. Guides are guiding him to you, too.

Why am I unconsciously sending my face to him?

You recognize who he is. Need to make it conscious for him to choose. More like him are coming. You're his teacher.

All kids are cute. I see you all the time. Torg, Ryxi, Kelly....pretty funny! Kids!

Everything seems like a paradox.

It's the nature of things. Yin/Yang. Each has the seed of the other.

Get a purple feather for your hair.

Tomorrow, make cake.

Dance, smudge, sing, teach mysteries, learn success.

Anything you don't want to tell us?

Names of all the stars.

Who is Shiva?

Angel of India. Not male or female.

July 10, 2011

Hello

Is this Heavenly Mother?

You bet! Blue hair now. Nice sweat lodge!

You were there with us?

Of course. Lots of visions. You and Torg set the stage well for visions.

You kids rock!

There are lots of saints and Gurus in history, I found a list on the internet, what do you think about some of them?

Many of these people do not keep coming back. You should think of these people as types, or animal totems, not individuals. Many of these people are not unique. There are many equal or better. They have mostly moved on.

Jesus-

My beloved Son does what I say...usually.

Babaji-

Not bad, smart. He doesn't keep coming back. New person every time. Great spiritual teacher.

What were his best teachings?

Love all, accept all.

Afra-

Not a master, good guy.

Lady Nada-

Taught peace, the same as 1,000 others.

Mother Mary-

She accepted her calling. Chosen because she was a good mother and clean. She was the best choice.

Hilarian-

Famous, not special. A builder.

Maha Chohan-

Better than most. All good parents are as deserving.

Enoch-

Better than most. Accepted his calling and gifts.

Confucius-

Wise, His sayings are famous and true. Listened to the spirit. He also came up with most of them on his own.

Elijah-

Many great deeds. Mixed. Angry.

Like me?

You teach well when you are calm.

St. Teresa of Avilia-

Most spiritually advanced of all my daughters.

Rowena-

Good witch. Healer of both physical and spiritual.

Mary Magdalene-

Mixed. Mostly good. Wife of Jesus, helpmate. Hard job. Many 1,000 women as good.

What is the difference between the spiritually advanced masters and the list You gave me of the top people on earth that have done the most good?

There were many masters left off that list. If you were alone on an Island you could become advanced, ascended, and not do any good. One way is to teach, and My people become advanced too.

Any other examples?

It's all about space and time, politics, many factors. A father nurturing his dying family through the plague is more deserving than a Guru on a mountain.

If he were able to help more people, would that be doing more good?

Of course, but he gave all. You were a single Mom in a bad space, you did very well.

I think I could have done better, but thanks.

Everyone can. You need to know who you are. If you stay with your calling, you will be greater than most on the list. No cutting yourself down anymore. It insults your calling and Me. Stay in balance.

I feel bad about not being able to pay all my bills. I feel irresponsible!

Jesus didn't. Mary paid the bills.

I am so different from my family and a lot of people who live around me. I feel I am misunderstood and disrespected for who I am.

The world hated all these people. It was hard for them too. It killed many of them. The Grid work you do is more powerful than the books. Your team is the most powerful Grid team. Yours is the only one with more than one onyx from Giza. Many people write books about Mom, very few know the elements. The books are your highest calling.

July 25, 2011 or 1012

Sorry I'm late.

No worries.

You're not mad?

Israel is thousands of years late. They haven't acknowledged Jesus for 3,000 years.

Do you ever get sick and tired?

Not for a long time. Dad is patient. It made him a God fast.

So all this patience I'm trying to have with friends, family and renters is helping me become a Goddess?

Yes.

When does patience become enabling?

Never. Be patient while I paddle your butt! We waited for the right time for the flood, wasn't enabling. It's almost time again.

This time with fire?

Some, kundalini spank.

Kundalini doesn't spank, does it?

Oh, yes it does. Kundalini is not a noun! It describes the times you live in. Earth, water, fire, air. All are very active.

So kundalini can be used in different ways?

Yes, think of an orgasm.

So it's explosive?

Yes, apply to the earth.

So it will be pleasurable for some people?
Of course! Some like it rough.

Does Dad know you talk like this?
Of course!

What do we need to do to make it pleasurable?
Get ready. Ready, set, go! Feeds on energy. Each round stronger or tighter. Some women climax during birth. Makes it more fun. Otherwise no more kids. Who would do that twice if it wasn't fun? It used to be much easier. Sing, water, good pizza.

Anything else on the horizon?
Clouds. Metaphorically.

What kind of clouds?
War.

In the U.S.?
Of course.

With whom?
You know no name. You know the "Pogo Principle"? "We have met the enemy and the enemy is us." You fight You.

Races?
Yes. Liberal vs conservatives. Gangs. Male vs female not as much. Started years ago. You are fixing that.

The book hasn't come out. How am I fixing that?
We've had this talk.

I think she means the energy acting to create ripples that change the physical? She must be talking about the male female grid ceremonies.

Was the Norwegian killer claiming to be a Knight Templar operating on his own or was it a set up?
He wants to revive the Knights Templar. He acted alone.

Why would this work?
People want power. The Masons are mad at him. He acted too soon.

If he had acted earlier, it would have been OK?
Top leaders of Masons have an agenda.

What is their agenda?
World domination.

Is this the Illuminati?
All are webbed together. Rothschilds, banks, Bilderbergers. The Illuminati is only one leg. Satan rules all of them.

Were the Masons originally good?
Tried to save Solomon's religion. So yes, they were good. Illuminati the same. Money the same. None were evil at the start. Bad Boy Satan.

My hair is striped yellow.

A God can make it work, Bye.

Aug. 8, 2011

Elements: *Things are moving all around Gaia. All prophecies are coming around now.*

Which ones?

Politics, weather, volcanoes, etc. You're in a bad mood, you're a Shaman, you need to retreat.

I had complained all week about feeling sick, low on funds, car problems, the usual!

I know, sorry I'm such a whiner.

We've seen much worse in history. You should have heard Enoch complain!

More than me?
YES! For 200 years! Mrs. Noah was a royal pain! Mary whined too.

Which one?
All. Mary Magdalene wanted Jesus home more.

Who was the worst whiner?

Sarah, had to wait for a baby. Ruth, no money. Had to gather food, God made sure she had enough.

Mom: *Hi kids.*

Sorry I'm having a pity party today!
I've had a few too! Don't catch me on a bad day, makes PMS seem happy!

I thought you were perfect and didn't have bad days.
Emotions are eternal.

I'm surprised you want me to work on my anger so much if emotions are eternal.
Mine too, you're in GOD company, Mine!

I'm sure others have more to whine about than me.
Not really, comes with the mission. If only the rich brats knew. Go watch a funny movie. Go laugh. Let's talk tomorrow.

Aug. 22, 2011

Ryxi, Love you

Who's here, Mom?
Me and others. A team. Your ancestors are some of your team members.

Maxine: *We are doing very well. We are leaving on an assignment soon.*

I need to finish the book on herbs with you. Sorry, I've been so busy!
It's last on the list for a while. I'll be back

Where are you going?
Another reality.

So does that mean you are resurrected?
It's happening now. It feels weird, but I love it! I'm getting more solid all the time.

How fast in our time?
Maybe a month? It's like assembling a team.

Are you aware of the different intelligences assembling

your body?

> *Close. Some are very fast like light imploding (seconds).*

Do you get to choose what you look like this time around?

> *Sort of, ask MOM.*

Can you choose to have green eyes today?
> *Mine are.*

Did you choose it?
> *Yes.*

Can you choose a different color later?
> *Yes. Wednesday.*

Who's going with you?
> *My husband.*

You got married?!!

> *Yes, we have to be married to go, Baldor is waiting! LOVE YOU! Bye*

> Mom: *None of you are a pure race. You have yellow. No one is pure. Love you kids. You're fun to watch. Like cowboys and Indians.*

Mission?
> *Glory*

What is your definition of glory again?
> *The increase and progression of the family.*

I think of it as Great, credit, glowing...
> *Glory is over rated. My children's glory is my glory.*

My guides and the elements say when I am sleeping ... I, them and Mom talk a lot. They told me we all decided last month to talk about the earth Grid, resurrection, volcanoes, etc. Many of us go lots of places when we are sleeping. They say it's like the movie Avatar, only we forget when we wake up.

Aug. 8, 2011

I was cranky with the elements today and they said...
> *You're in a bad mood, can't talk clear now.*

116

Aug. 2011

Hi kids. The new Millennium will be about opposites coming together. An Icy flame either purifies or destroys.

What age do you look like?
Between 20's and 30's.

Aug. 30, 2011

Over the weekend we went to Safe Haven and with instructions from our guides we closed an old portal across the valley from Safe Haven. We were told an evil witch lived there at one time and opened an evil portal and that it was still open and affecting the energy at Safe Haven, as well as the surrounding area. She did ceremonies and sacrificed children for magic, Ughh!

Gaia is Happy you closed the evil portal. Did well! Forgot chocolate though. Thanks for clearing garbage from the mountain.

Is the Portal closed?
Mostly, fine now.

Did the evil woman live there?
Some of the time. During ceremonies.

Did she steal the children?
People brought their own children up to the mountain to be sacrificed.

WHY?!!!
Power, of course.

Did the babies feel pain?
No. Most peoples had similar practices (cultures & races).

What is the difference between Abraham and Isaac's sacrifice and theirs?
We stopped him. He knew we could bring Issac back. Test of faith. We did the same with Jesus, tested his faith.

Torg's had his faith tested!

You too!

How?

10 years without a 9-5 job and you haven't lost your house. You're still working on faith with the book. It will never be finished. Ripples.

My hair is white, tried Rainbow.

Will we be going to Hawaii to work on the nexus there?
Plan for it.

Why does it say to kill witches in the Bible?
Not witches like you.

Evil Witches?

Yes. Words don't mean the same now as they did then. We meant people fighting us, claiming power. Moses was a great magician.

They don't call him a magician in the Bible.
My point. Language was different then.

Anything else?
I'm wearing pants.

What color?
Light blue tunic. Long white hair and barefoot.

Anything around your neck?
Not now.

Wow, sounds beautiful! We are tired, can we talk another time this week besides Sunday?
Maybe.

Love you, Thank you!

?????

Shamans work with animal guides, what do You think of animal guides?

All plants and animals have defining powers.

OK to call on those powers? Usually metaphors?
A mix. Torg's beaver is healing, metaphorical. Dragon is personal. Reality.

What is my Power animal?

When? It changes.

.. **Now?**

Female moose. Strong Mother. Don't get a Moose mad! White buffalo calf during Jesus phase. Still call if need.

Jesus phase?

Accepted his baptism and new name.

I also had a vision that Jesus baptized and blessed me. I didn't understand why he baptized me because I was already baptized. He told me I was an independent contractor. I also saw a vision of myself riding on the tail gate of a pickup truck and a white buffalo calf was chasing us in fun. I was told by my guides it meant I had to do my job before Jesus could come. The white buffalo calf represented Jesus.

I was told I needed to do my job before he came...Have I been doing my job?

Yes, His part. You are mine now.

What was my job?

Bless Gaia. You're doing that.

Why Baptize me?

Needed to take His name.

Didn't I already do that?

When you were younger, a reminder.

Was his blessing just for the Gaia calling?

ALL. Look at the John scripture. John 3:16. On the "In and Out Burger "cup. Jesus' brother.

Love your mother-in-law. I Love her, she's Mine, soon will come home. Your Mom too. You will miss them.

Go rest, bathe, have fun. You did good the last two days...Book, Study, etc.

Sept. 9, 2011

My heart seems closed, I don't cry much.

You don't want to get burned. You feel you need to be tough. It's hard to love that way. If your soul isn't open, you can't cry.

119

Do I need to open my soul?

You would get hurt. Are you willing to get hurt? You also get Love.

The Shaman path is usually poor until they get strong. Have to stand alone to get strength and power.

Remember, you still work for Father, Jesus and Holy Ghost. You just tell My story.

Sept. 12, 2011

Who's here?

Lots.

Mom?

Yes.

Thought so.

White Hair, Blue sandals, exquisite fish scale Toga.

If you are wearing a Toga, your shoulders are bare, isn't that against the law? Shouldn't shoulders be covered?

I'm in my home. Not strict on that. Strict on butt cheeks showing! Not showing much, some cleavage.

What age do you look like?

Pisces.

Avoiding question?

Torg's too old for me, I look 27. I look like Torg's Gaia? (A picture of Gaia Torg has in his office) I look like late 20's, early 30's.

Is that a picture of you?

Yes, I do a mean sexy dance for father!

Teach truth, no conflict, be a scholar.

Sept. 14, 2011

I hope I thank you enough.

Not bad, You should have seen Noah.

I just want you to know I am grateful.

I know. Teach mysteries, you do well.

Sept. 15, 2011

I love Bob Marley's music.

I liked him too. He was foreordained like you. Not all of what he did was foreordained. Had unclean hair and bugs.

That matters?

Yes, long talk. He left the path for a while. You study him and report back.

He was an important part of his time?

Yes, like you. You're a long talk. What do I tell you to do all the time?

Dance, sing, teach mysteries and listen to spirit.

I told him the same thing.

Do You want me to write and sing?

You do all!

I'm too old.

Bull!

I sing to the Grid, not to people.

You sing to the world!

The Grid I understand, do You want me to sing to people in the world too?

Why weaken the message? You sing to the Grid and sing to Millions! Huge ripples!

Mine is subtle.

No, Tsunami!!!

Don't You understand why I don't see the effects? As far as I can tell I'm just singing to myself!

You do, too, see!

I guess I'll take Your word for it and not argue.

Good! Madagascar, Pele, Picci, Mesa Verde...getting the point?!!!

I really don't see what's happening.

Will you open your eyes, please!!! You manifest everywhere! It's scary!

Why is it scary?

This is your calling! You don't manifest everything, but most!

So I'm a very powerful manifester? I have that gift?

It's ABOUT TIME!!! You need to study the basics like chemistry, physics etc. so you are not so frustrated. Become a scholar. Learn how the world works. You will enjoy more when you know more. Many things you miss.

So I'm a Master Manifester!

M&M's!

Do You like M&M's? (The candy.)

Yes.

How do You get them?

Not hard, In stores all over. I come a lot. I dance and play with my kids.

Have I seen You?

You saw Me, but you haven't seen Me.

Do very many of your kids see You?

As I am? No. It interferes with agency when you see me as I am.

Because they wouldn't have faith anymore, they would "know"?

Very good, pat on butt!

Why is it so important?

It's not that simple. Another very long talk.

I heard the Natives are gathering at the Vatican to deliver all their records.

Some, Yes. Prophecy. They will go to other churches too.

I wondered if their records are like the Nemenhah records that tell of Jesus visiting them in Sanpete County and upstate New York after He was resurrected.

122

The Hopi have legends that a White God came to Second Mesa, taught, and then walked off the Mesa into the sky.

The Mayans have a similar story about a White bearded God coming to them and teaching them before taking off into the sky. They call him Quetzalcoatl.

Our guides have said that there were many places Jesus visited and that records of His visits would come out eventually.   Torg has a friend who has many friends from different Indigenous tribes who have records of Jesus and that they are getting ready to bring them out of hiding. Cool!!!

Won't it be dangerous for the truth to come out? It will shatter paradigms.

Yes.

What will the Vatican do with them?

Keep them.

Will they share them?

Hope so.

Why the Vatican?

They are taking them to all peoples and places: Chinese, Greek, Mormons, Buddhists, Arabs etc. The fullness of times. Information not lost. Share all, no more secrets! The flood lost much, not again!

So when this stuff happens, we won't lose info this time?

No, Torg knows. It is that which it is, is it not?

Sept. 27, 2011

Torg and I had been working for the last few months on bringing a Global festival called Earthdance to Salt Lake City. There are simultaneous festivals going on all around the world at the same time where everyone prays for world peace and the healing of the Earth at the exact same time. One year we used the energy of everyone praying all the same time to create a new node that was missing in the Caribbean from when Enoch left the planet. This was our sixth year and it doesn't get any easier. It is very stressful and hard work. We were exhausted.

Hi kids, I'm proud of the festival. You've been working really hard on the book.
Just organizing.
Hardest part. Good festival, take a break. I will assign guards here soon.
I thought we already had guards.
Shields yes, Power yes. No exchange guards yet.
What?
Other universe guards.
Why from another universe?
Give them a job. They report back. Some are over the Pacific with nothing to do. Re-assigned to your team. Your team is getting strong. Attracting attention.
What do they look like?
Spinning wind maybe?
How many teams like ours are around the world?
Less than 20.
Thought there would be more.
Strength matters not numbers.
More strong than ours?
Yes, why do you care?
Just wondering.
When you do your job, you will meet others. Some you already met. Isis, Madagascar. Four teams.
I thought we were all one?
True dat! Human teams. Africa, Australia, Iceland. In Africa the team was physically there. Earth entry, power Grid. People together in oneness. Things all over affected. Very long talk. A baby in China stopped crying. A storm in the Andes turned away. A penguin's foot healed. The drum worked. Yin/yang too much work.
How do we make it more powerful?
More prayers.
Why am I so exhausted?
Torg is right, Gaia is twisted. You feel it. The food you're making will help you sleep. Gaia is twisting under us.
Will there be an earthquake here? (Salt Lake City is on a

124

fault line.)

Not soon.

Sept. 28,2011

I have bare feet. I have a tree of life tattoo.
I thought we weren't supposed to have tattoos.
You, probably not. Mine comes off.
When we pray to Father do we get the same answers as when we pray to you?
Not much difference. We are one...mostly. We talked about this before. Pray to the whole Family, let "Them" give the best answer as a whole. Like putting heads together for the best answer. I have the best head.
Many people live the gospel of peace and are Christlike but don't follow Christ.
Normal. All the prophets had failings. Your friend is a good worker, mechanic and Shaman but sleeps around. Same with you Ryxi. You still have doubts and get angry.
Is mine as bad as sleeping around?
It affects your mission. Sleeping around affects salvation.
I can still finish my mission with my doubts...
Correct. Hard, but, yes. You can repent of your doubts and still finish your calling.
My toes are green.
Are you at home?
Yes, in a robe. Might change my tattoo to star of David.
It seems weird that sometimes when I pray for help a drunk person shows up. I thought it was hard to listen to you when people are drunk?
They show up to help Me so I can help them. Wheels within wheels. All are rewarded by deeds.

What about accepting Christ?
Tied together, a package.
What do you think of eastern religions, like meditating

and Deeksha?

Deeksha works. Would work better if Christ were in the package.

What are they doing?

Chi is spirit. If they prayed before working in chi fields it would be better. All professions are better with Jesus.

October 2, 2011 (Mom)

Hi, how are You?

Godly.

There is a lot of controversy about gays and lesbians in the press right now. Conservative religions are very strict about it.

A strong stance will help some and We will help the others. Many are on the fence. A strong stance will make the ones of the fence choose straight. Those who are naturally gay, We take care of. Service is Godhood. They serve well. No ambiguous Gods.

The Bible says no gay.

Yes.

Was that supposed to be true?

General. Can't include all special cases. That's for personal revelation.

Seems it would be easy to add in the Bible, "Except in special cases".

People would look for loopholes. No kill unless need. No adultery unless you're in love. No covet unless you need it. Honor parents unless you disagree.

Are you saying gays/lesbians can't have sex in your kingdom with each other?

No, they don't want to. Forces are not aligned.

I can't imagine not wanting sex!

No way to give you understanding. Sorry.

I really want to get this book finished.

Long time yet. Too early to worry.

Here I thought I needed to hurry!

Still a year or more. You no book until scholar.

Don't know why. Seems like being innocent would make more credible.

Can't be safe in ignorance.

I wanted to "be a scholar" on the subject of YOU. There's not anything in the Bible about You.

No, it was written by men. Find other sources.

Anything in the Koran?

All over.

Dead Sea scrolls?

Yes. Learn diplomacy. Not just about Me. How to talk, write, present, debate. No conflict. That's why no anger. If you say it bad, it will ruin the message.

How do I learn this skill?

Council from experts and friends. Avoid triggers. No I'm right, you're wrong.

It seems I should have the talent for this if this is my calling.

You do, you just let anger cloud it. Your job is to help people grow, not rub their noses in it.

Do You play games?

Of course.

Any we know?

Sex. Want to know what I am wearing?

Sure.

Won't say.

Nothing?

None of your business.

Be glad Rama's not your mama.

Why?

No fun. Hates reggae.

You need to repent.

Why?

Bad hair day. Still angry and irritable. You don't listen when you are irritable. I get irritable too, but I still listen.

Who do you have to listen to?

I have red toes and a black stake in My hair.

Are You copying me? *(I had colored my hair)*

You are copying Me! No nose ring.

I have a planet like Hawaii. You can make one too, if you behave. LOL.

October ?

What is the best way to fight Satan?

Information. Light and dark don't mix. Become scholars. There is no saving in ignorance.

What do you think of the Occupy movement?

Educated people are not getting involved enough. They don't have a clear purpose. The return of sovereignty is clear. For government, corporations, and schools.

We need to be specific and clear about each one?

All take away agency. God's laws make people free if they study them.

For example, laws to stop at stop lights; is that good for agency?

Yes, it frees people to be safe on the roads. Agency infringement would be to not allow people to speak truth in public. The 10 commandments all improve agency.

What about the law to not steal, this removes agency.

The big picture, Ryxi. All laws are supposed to protect agency. Most do not. Licensing only protects the rich. Patents can do either. Big corporations use patents to stop others, like corn genes. Can't patent corn, it's yours. Not corn, but new genes. Yes.

I should go through laws with You and get Your take on all of them.

Yes, a very long talk. Laws that make governments big are usually bad.

128

Which laws right now do we need to get rid of? What is Your suggestion for a plan of attack.

Education. Fix schools and teach people to think. Put parents back in charge. Put Us back in schools!

They say it is a violation of church and state.

Nonsense. We wrote that. One way only. The state can't create religion, but the religion has to create the state!

Will the Occupation movement help?

Good idea. Tried many times before. Many more to come. Torg's been there. Like Gandhi, need to help educate others. He followed the spirit. Maybe call them "out of control capitalism".

What is the best: communism, capitalism, socialism?

They mutate, I don't like "isms".

What should we replace them with?

Sovereign individuals. We need government to protect sovereignty. Individual responsibility.

Who takes care of someone who gets sick and can't work?

Community. The Essenes tried; not function tor a long time because of enemies. Athens was good except for slavery.

So You didn't protect the Essenes?

Agency for both. We helped but they stopped protecting themselves. Slaves could be free. Free could be slaves. Like now, some work for selves, others work for someone else. There would only be slaves if a crime is committed or they go into debt. Slaves that were bought could buy freedom.

How could they buy freedom with no money?

Only some of their work was as a slave; also worked overtime and side jobs. Many old societies made everyone equal - aborigines.

Why?

Few laws, all sovereign. Few power hungry. Few possessions.

What kept them in check?

Education.

The Hopi say they have no possessions because God said no. What about aborigines?

Yes, educate about Us in schools. Your laws don't say that. Don't covet. Understand spirit.

World Trade Center wasn't Al Qaeda; Luciferians. Bush.

Maurice Strong?

High up.

Is Evelyn D. Rothschild on top?

No, never was.

Who is at the top?

No top; coalition of Chinese and Bankers, scientists, broadcasters, politicians, rulers.

What is their motivation?

Power, sex. Earth Gods.

What do we do?

Educate, spread light.

Should the book come out in a year?

No hurry; scholar first. If you want to be safe, you can't be ignorant. You go tomorrow and protest with occupy.

I found out later that the occupy movement would be put into the "liberal" category and become part of the republican vs democrat fiasco. Satan will either try to twist a movement or find a way to use it for his agenda. I'm so glad I am surrounded by friends who are issue oriented instead of following the crowd. One of Satan's best tools. My friends rallied with the tea party AND the Occupy movements.

What do You want me to study?

All: history, science.

130

Oct. 8, 2011

We had our first Goddess workshop with our Goddess team. I was really nervous to share about Heavenly Mother because most of the people there were from conservative religious backgrounds. I also taught about Gaia. Even though She is an out of the box topic for a conservative crowd, I had taught about Her many times before so I wasn't too nervous.

Great job Goddesses. I'm proud. You showed your stuff!

Showed my stuff?

Even though you were nervous, you did well. You strut well!

Feedback?

Need more time for questions. First time big talk. How do you think it went?

Not sure I'll share the same things next time. Not sure people are ready for your sense of humor.

Too used to "fire and brimstone" Gods.

What about fire and brimstone?

People twisted it. We get to have fun too!

Did I do what you wanted?

Yes, Yes, Yes! Demonstrating blessings the best! Do again. Ask questions and ask input.

Rider do this again?

Yes, her calling. Ripples build, time short! Finesse presentation. Visual aides well done. Do more.

What visual aides?

Gaia ball, blessings. Show a combined blessing with men. Yin and Yang. This is a dry run. Next time more on child raising and sharing roles with men. A whole conference on me! Be careful and smudge.

Make a manual for reference. Speak only truth.

Did I not?

Did well for your first time.

Are you saying I was frustrated and angry?

Not really. Always can improve. Use scriptures more. People need to get out of the box gently. Listen to Me always and make your own way. The spirit was there all the time.

131

I'm only a bad diplomat when I'm frustrated and angry.

When you need it the most. Teachers for diplomacy are Gandhi and Jesus. Part of skill, know when to make a whip! Sometimes can't be diplomatic. Mother Teresa a good diplomacy teacher too.

I'm all white now.

If I learn diplomacy will it bring generational healing to my ancestors?

Yes, all the time work on yourself.

How?

Pray and ceremony.

Isn't that true for everyone?

No, sometimes makes worse. For instance, your mom tries to anchor you to her when she prays. She tries to make you follow her law. Needs to break connection, not make stronger.

She prays to make connection stronger so I will follow her laws?

She prays, "Please make my kids listen." Parents should pray that their kids live well, not make them listen to them.

She thinks if we listen to her we will live well.

Family ties are the strongest when the bonds are broken. You know the story of the eagle? "Let your eagle go, if it comes back to you it is yours. If not, it never was." Give council carefully, only when asked.

That's a hard one.

True dat!

I don't cry.

You get angry. You need a role model or a spanking. You, Rider, and Kelly are Goddesses. Are doing better but can still improve. Torg's hopeless. (Teasing.) It's critical that Dorothy grow or she'll die. Both needed to grow up!

You are doing better.

Is my anger issue generational?

Your Mom didn't have it. In your genes, yes. Luck of the draw. A predisposition. Like Torg's heart and diabetes.

What can I do about it?

Develop strategies, be a diplomat, have friends that will spank.

I can't get rid of it?

No. Control it, yes. I'm proud of you for how far you've come.

I will have blue hair tomorrow.

Why tomorrow?

Cause it's not today.

Do you have an appointment with the hairdresser?

No, I'm God.

So you just tell your hair to turn blue?

Sort of.

Have you always been this humorous, or does it come with being a God?

I'm addicted to humor.

Do most humans have demons (addictions) to overcome?

Most humans are demons. The natural man is an enemy to God.

Love you, teach mysteries.

Oct. 10, 2011

Are you sick?

Yes.

Take garlic, thieves, body work, Matrix and no walk today! You need pampering, no whimpering! No brat. I'll have to spank!

I've been laughing, doesn't that count?

You're fine so far. Don't lose it!

Did you send the angels I could feel blessing me last night?

Three worked on you in a triangle. Yes.

After Matrix I felt a sensation over my crown chakra. It felt like I was getting a Deeksha blessing.

Thanks!

You're welcome.

133

Nice pumpkin on the porch.

Call your friend Travis in Pele (Hawaii) *and let him know you are coming.*

I feel like the angels are still working on me.

Yes, you should go to sleep.

What are they working on?

General. I have blue hair.

Is the "occupy" movement working ?

Will improve agency. Mostly good. Satan will try to use it all he can.

Like how he used sex. Sex can be used for both good and evil.

He turns things evil like money, etc. Smart Ryxi, you see evil and good.

Can't most people?

No, that's why it works for him. All can be used for good and twisted easily. Guilt too.

Example?

You feel guilty and think you're a bad mom. This is Satan speaking!

Yes, but if I hadn't....

STOP! If Satan knows you feel bad, you are not as powerful! Torg passed test with ego, no worries. Both stop, move forward. Rider feel insecure, go figure! Kelly is locked in old patterns, needs to get out! We need you all, get a grip!!!

You four are powerful! (Me, Torg, Kelly and Rider.) *You really have no idea! The Goddess weekend didn't teach you?*

Yeah, kind of.

Get it figured out! Stop, act your power/dance.

You three and Torg will be called to act many times! You must be sure. You must cover each others' backs! Your strength is greater than any but a few.

I tried to find a way to change the feeling of being overwhelmed by it all so decided to try to look at it as a fun game.

It will be fun! We are four ninjas! What makes us powerful?

You don't know?! You studied and trained a long time before for this purpose. Even the cat! Doesn't it seem odd she is so young and yet so old? This is your time. Study and be a scholar! Prepare to fight Satan and meet Jesus! You've been in it a long time now! Hilti and Milni assigned to you before you were born!

Pongust to Torg?

Many to both of you!

Who are the few more powerful? I would think Enoch and Adam....

Moses and Jesus.

Who else?

None of your business. Your real challenge will come soon.

Can you tell me what that is?

No, scholar. Let go of your fears of knowledge. No time for that crap!

Why am I afraid of knowledge?

You think you can't learn stuff.

The only thing I'm afraid of is science, physics, etc.

False fears. Can't be safe in ignorance.

What shall I start with?

History of Egyptians, chemistry and physics. Torg tutor. The new world needs scholars.

Can't we just share all the books? Do I have to learn all this right now?

Tomorrow.

Do I need to get a degree?

No

I would think my biggest challenge would be to stand up and talk about things that are too out of the box.

Minor! You fight Satan for the Grid!!!!

I will save the earth from Satan?

You already have! You know this! Madagascar, Giza, Mesa Verde, lodge, Goddesses. You are fighting Satan and saving the earth!

A few years back, we were asked to go out of body with a team of Shamans to Madagascar to take the node there back from Satan. Some were even animal Shamans! There was a shaman parot, elephant, squid, turtle, whale and small life (organisms).

What if I were to quit, what would happen?

Someone else would take over. You are the best choice. You are not essential, but you are best. You don't fail!!! Jesus didn't! You are his sister, act like it!!!!

You are a God, go do God stuff!

Is more money coming so I have time to study?

You have tenants. You are learning management skills, part of being a scholar.

So, I'm not doing enough scholar stuff even though I'm learning management skills?

No one does.

Do you want us to meet with you before Sunday?

Yes, anytime. This is your shift time. Try not to ask any blonde questions. I'm not going to be offended if you get angry with me. Try not to have an ego. You're fine so far.

Oct. 18, 2011

We took some of our Shaman students to Safe Haven for a vision quest. I decided to do my own quest and Torg guarded for all of us. Safe Haven is on the edge of a large hill made of almost white dirt. On Google maps it is called White Hill and looks like a dragon. I climbed to the top and as I climbed I felt I was climbing Mt. Sinai! I also take people on a journey during sweat lodges to a place called "sacred mountain" to heal and see visions. I camped three nights by myself on the hill with my books, notebooks, fruit and water. At night as the sun set, I would play my drum while looking over Safe Haven. Torg and our students said it was surreal to see my silhouette against the setting sun and hear the sound of drumming. I received some cool visions. After the quest we had a sweat lodge and feast.

Your weekend went very well. Lots of closure and lots of new beginnings. Loved the lodge. All had revelations.

136

I'd like to know if I got the visions right. What came from You and what from me?

Not just Me or you. Came from the guard, elements, Me, you, ancestors and friends....

What one came from You?

Too long of a talk. It's time to learn new skills. Torg might be leaving soon. You need to be independent.

OK, how about verifying one vision, the beginning of the creation of the universe?

You tell me what you saw and I'll tell you what I sent.

I want you to tell me first so I can verify.

I'll try your way first. I showed you how it happened when we spoke the word and the universe came forth. The important part was our job in the creation. You probably weren't able to see the whole creation accurately, it was just too huge. Probably looked like sparkles in all directions. Probably like kids let out from school or prisoners set free. Their joy and fear matched ours. How did you feel about your kids leaving home? Fear and excitement.

I saw sparkles coming through and saw your joy when you saw us!

You didn't pick up on joy and pain? I wanted you to "feel" more than "see". I think you are more capable of feeling than seeing. More than my emotions, science. Were you scared for your kids to go to school?

I don't remember being scared. I guess I was worried they would do well and fit in because they are different, mixed race.

Most of you kids have a very hard time.

Oct. 24, 2011

We were finally able to go to Hawaii to work on the node there. Pele (the volcano intelligence/power earth element) had been begging us to come for a few years and we finally had the funds to go. She said she was barely hanging on. She was afraid a part of her would fall into the ocean and create a tsunami that would devastate the West Coast of the U.S.

We were instructed to meet and receive a blessing and a new name for me from a Hawaiian Shaman (Kahuna), do ceremony, and Torg was to meet the other dimensional beings guarding Pele.

I was hoping to meet up with the whale I ride sometimes out of body, who was upset I hadn't been able to meet her in person yet. They said she was still feeding in Alaska, but I hoped to meet other whales that may have migrated back to Hawaii early.

We took one of our Shaman and Goddess team members, Kelly, with us and had a blast!! We stayed with friends who had helped us start Safe Haven, then got married and moved to Hawaii; how convenient! God works in many mysterious ways.

Nov. 16, 2011
Good job, you are a star in our universe. You found everything but whales. Next time.
We are going again?
Of course. You are going to build a safe place.
In time? For what?
Like Safe Haven, for changes. Long time to change. There will be many places. Pahia was the first choice. Her mom liked Torg. She felt the power of three. Pahia can now work in the open. You started ripples that will affect all the islands. You had more of an effect on the islands than they had on you.

Pahia was the Shaman we ended up meeting. Our time was almost up when we finally met her. I was beginning to wonder! When I told her what we were there for, she got what she called "chicken skin". She said we were "Children of the Rainbow".

She took us to an earth portal. Hawaiians call them Heiaus. Her expertise is doing ceremonies at different Heiaus around the island. Go figure! Before she gave me a blessing I was wondering if they had a legend that Jesus visited there and if she believed there was a Mother God. When she gave me my blessing she started her prayer with, "Dear Mother and Father God". Wow! Then as she blessed me she told the story of when she was in ceremony one time and God told her their legends of a white God who visited Hawaii long ago was Jesus Christ. I was floored! I forgot to ask for a name but later when she gave us a ride to the airport I told her I was supposed to get a name from her and she turned to me and said," Your name is Whale Rider." Wow again!

She shared with us that her family didn't understand her calling to be a Kahuna and felt supported that we were there confirming her calling. She asked if the three of us could give her Mom a blessing together, the "power of three".

She took care of us and gave us a ride to the airport. We had our car towed on accident and had to pay $350 for the car to be released. We didn't have enough money left over for food on the way home. I said a prayer asking for help. I was willing to use the experience as a fast if need be. To my surprise she turned to me and said, " I am supposed to take you to the store and buy you food for the trip home".

Thank you for all the miracles! It was a beautiful experience!

Thank you back! You did all I needed and more. Big rock in big water.(The intelligence over the Pacific ocean calls himself "Big Water")

I wish I could see the effect. I just played in the waves.

It was just play, down time, when you were in the ocean (Big Water).

What was the work we did? It all seemed like play to me!

Time with Kahuna, Livingston, the new information you gave to the guards. They are more focused now. Grow good plants.

What new information did they get?

More understanding about the way they are used... Food, shelter, pretty, harmony, etc.

Should we still be teaching Shamans?

Yes. Teach in Hawaii. Go there.

Have Pahia come here?

Good Ryxi! Always more respect when from out of town.

Australia next. Go meet the wise man and the wise man's partner. You teach, they teach.

Is this the one I met in a vision from New Zealand?

Timing. You'll meet the right one. Probably do

139

ceremony, blow conch. Find Shaman to teach and be taught. Study first. Study, be a scholar.

What were the most powerful and important parts of our trip in order?

1)Torg meeting the Guards. 2) Kahuna 3) Torg's dance. (At the Polynesian cultural center on Samoan stage, silly!)

Thanks for the airport finale!

Was not Me, you were in the flow.

When I got off the airplane back home, I heard the conch shell being blown in the airport and at first I thought I was hearing things, then I heard Polynesian chants and saw a group of Polynesians singing and dancing, welcoming home a missionary. I was so touched, I cried. I felt God was welcoming me home too. How could it be anything less than a miracle, what are the chances?

The conch shell itself was an omen. I had first heard it being blown by Chumash Natives on the beaches of Malibu, California. When they blew the shell, tingles filled my body. It was like the Holy Spirit blew through the shell. I knew I had to have one. I had tried to find one. Torg looked on line for one to give me for my birthday and they were too expensive. While in a gift shop in Hawaii I suddenly thought of the conch and wondered if there was one in the shop. My body turned automatically and I was guided to look on a bottom shelf and there it was, a conch shell ready to blow! For a good price too!

Nov. 17, 2011

Where are the guards in Hawaii from?

You don't have enough background to understand.

On exchange?

No, many places, times and entities. Like your time dollars. (A time bank we started.)

A different universe?

Close.

How was Torg able to communicate back?

Many Kahunas have talked to them. Had to think in Kahuna. Language of Kahuna.

Words?

The show at the Polynesian cultural center helped.
Good Torg.

We watched a powerful spiritual presentation called "The Breath of Life" at the Polynesian Cultural Center that made me cry.

You are welcome.
For what?
Livingston. Broken fuse on the plane movie system so Torg and he could talk. Many other people talked. Romances. Ripples!!! So much fun with blowing one fuse!

The first miracle of the journey was when Torg got on the airplane and they put him at the emergency exit. During the flight a man asked to sit by him because of the leg room.

During the flight the movie wouldn't work and they got to talk the whole time. It turns out he worked at the Polynesian Cultural Center and used to be my ex-husband's boss when he went to school there! My kids were thinking of going there and working at the Polynesian cultural center! Freaky!

How did you put Torg and Livingston together?
Nudge. Kept others from that seat.
How did You get them on the same flight? Good timing, right?
Only God knows
I just happened to book the right flight?
He had tickets a long time ago.
So did You know or was that random.
I have green nails and my lips are green. There are no accidents. You were In the flow and guided to that flight.
My hair is red.
Christmas colors. What are you wearing?
Not much.
Good to have you in a good mood.
Godly.
What is Masau's job? (Guard of Hopi)
To scare and protect.

What is the guard in Australia like?

Heritage. Keeps people away from sacred places.

What does it look like?

She, Ryxi knows.

I do?

Tied rainbow.

Torg saw a rainbow and a tied rainbow with me during a Matrix session before we went to Hawaii. I didn't know why he saw rainbows until I got off the airplane in Hawaii and then I saw them: rainbows everywhere! Even the car license plates had rainbows! The caption read, "The Rainbow State".

The rainbows there were the most breathtaking I'd ever seen. There were rainbows in the sky most of the time. Then Pahia told us we were Children of the Rainbow. It reminded me of all the rainbow prophecies of the Native Americans.

We found a book about Hawaiian spirituality while we were there that fell off the shelf! It explained the Hawaiians had a prophecy that a generation of people in all races would be born that would bring peace.

I thought the tied rainbow was about Hawaii?

Rain brush. It's brighter now that rainbow children are stronger since Ryxi came. Ripples. You can't drop a rock and not create ripples.

Seems silly, so my energy connecting to Pele was like a rock?

You always make ripples! You must understand this!!!

Physics?

YES! You can't move or stay still without ripples.

Was my intelligence that way from the Big Bang?

Yes! Noisy intelligence.

Does Torg make the same kind of "noise" I make?

Not that kind of noise. When you get mad, frustrated, sad or depressed, all things go down! When you are happy, all things stay positive. All goes up!

The world?

We are all one.

142

It's not just me, everyone?

A few like you. Whatever mood you are in more powerful, makes more ripples.

I thought You wanted me to feel all my emotions?

It's not about feeling. It's about acting. Your son is still alive cause you had Grid power to use. If you had been angry, he would have died.

My son had a close call and almost crashed his car. The elements helped by physically guiding his car away from danger. They used Grid energy to do it. They said because I help so much with the earth, I had a positive bank account with them and they were willing to help.

At that moment?

Part, a bigger issue. If your people want to have power, you need to be good. If you get mad, maybe won't be enough to save them next time.

My son?

Anyone!

Does it affect all, or my people specifically?

Both. You're not equal (balanced) when mad. Lose more Grid energy than when happy. Your nature. Others are different.

How many like me are on the planet right now?

Less than 100. (Wake things up and affect the Grid more) Grid tied power walkers. Pahia, your niece, Rider.

World leaders?

Dalai Lama.

How does the Rainbow Knot do her job?

Sits in brain and makes think.

How many Grid Tied Power Walkers are creating bad for the Grid?

Won't say. Nature, agency. When irritated, find better ways to act. Diplomacy! Don't let jerks win!

You make a rainbow knot?

What shall I study now?

Follow spirit.

You ARE the spirit!!!
> *I hire spirit.*

OK, I'll ask you!
> *You study science right now.*

What else should I do?
> *Go surfing, you need to have some fun!*

How is the Occupy movement doing?
> *They are finished.*

Is their job done?
> *Yes. That idea worked and is over.*

Did I go out of body in Hawaii?
> *Yes, 3 times. Looking for whales.*

Is that why my Hawaiian name is Whale rider?
> *Part, you were looking around.*

Rider was with us and asked to talk about her friends who were waking up and how surprised they were to learn new things....

> *A big move for your friends. Trust yes, not easy. Father Joseph is laughing. Doesn't do this often. He likes you (Rider) nice hair like his mom's, desert hair. (She has dreads) Step dad to God a big job!*

Joseph as in Jesus' step father?
> *Yes.*

Why is Joseph here?
> *He is an ancestor of your friends. Many are. Ryxi's kids are, too. Many of your friends' ancestors have been waiting a long time for him to wake up! They are excited. Good job! You part the waters of ignorance for them. Let my people go! You don't make Moses' mistake.*

Which one?
> *You speak, Moses wouldn't. No ego, no apologize. Acknowledge source.*

How are we doing in those areas?
> *Usually good. Every 28 days. Ryxi, you're doing good at no anger, be a scholar. It's hard, you keep doing better, I'm pleased. Do you want a number? 8. Ha ha.. Ego no problem yet. You're still apologizing. Not much though, you're doing*

fine. When you stop apologizing, you could get into your ego easier. Burn dinner, it will help to make up for ego.

Thank the lord.

You're welcome.

Sometimes I get mad at you. Did you ever yell at your parents?

Sure, no problem. You're in integrity. Tell the truth as you see it. Diplomacy would be nice.

I wanted the book to come out for Mother's Day!

No. You become a scholar first, teacher. Soon to be released. God only knows.

I was worried I was trying to push things too fast.

You're like me. Hunger close. Wow for their food.

November 28, 2011

Kelly was going through some hard times with her boyfriend and wanted some counseling from mom.

Hi daughters and Torg. Trials hurt and make stronger.

To Kelly- Friendship is stronger than arrows. (Romantic Love). Kelly having struggles with Hunter. He is still a kid. We make kids grow with trials. Spank and praise. You're a parent and a friend. Are you up to loving him enough and spank him too? You kids are fun. Been there. Can you love your friends and spank them too?

How can Kelly help?

Pray, Imagine that! Love always trumps and triumphs.

When You "were there", how did You get through it?

Pray, imagine that! Again, Love always trumps and triumphs!

Dec. 3, 2011

Elements: *You have lots to do.*

Top of the list?

Australia, Machu Picchu, Giza.

Stonehenge?

Others are more important. Not in order, go with spirit. Go with Kahuna, spirit. The spirit isn't one thing. It's a

145

combination of Heavenly Mother (GOD), going with the flow, us (elements), being a scholar, etc.

I'm including this next part because I want the readers of this book to understand how we manifest things in our lives.

I chose the date to go to Hawaii and then I chose the date to come home exactly when the missionary from Polynesia got home?

> *Smart choice (scholar), ended up with missionaries at the airport (go with flow), Heavenly Mother (God) opened the way and made you feel a part of the experience. It was a combination of forces working together and more. You are growing, only children let others take responsibility.*

It seems that if I'm given a job, the way is paved to do the job.

> *Job yes, the method is not firm. Do you make pie or cake first as long as the dinner is done on time.*

So like Thanksgiving dinner...The Turkey, dressing and pumpkin pie are necessary for a thanksgiving dinner. It doesn't matter what recipe or if the food was prepared the day before and needs to be done on time. Chocolate pie is an extra.

> Pongust said: *I am chocolate pie.* (More about Pongust and elements later.)
>
> *In the case of Hawaii, we asked, please come. You made it happen.*
>
> *You make decision, We and Heavenly Mother help if it is a good decision.*
>
> *You keep making good decisions and we will keep helping. Gone is the day We tell you what to do. You are grown up. You have new skills now. You just tap into the flow like Us. If you miss big, We will tell you.*

Dec. 6, 2011

The solstice was coming up, we've had major things happen on solstices of the past. My first winter solstice when I began this path

146

was in 2005 when I went to a Mayan temple in Guatemala and attended a Mayan ceremony with a Mayan priestess who told all of us we were Shamans and we would be helping with the shift.

Another year we were asked to take back a node, in Madagascar, that Satan had control over. That was crazy!!! A few years ago I went out of body to Stonehenge. I wondered what we would be doing this year. When we started our session the elements showed up first.

Elements: *This years' Winter solstice, stay simple. Candle lodge with a few people and Blow conch shell.*
That's it?
Plenty...next year OMG!!!!
Have I done my job from the Grand council meeting last year?
Extra!!! Mom and Dad are proud, we are blown away. No specific to Pacific, Kahuna will finish the process. You worked on the Pacific area when you went to Hawaii and visited the whales in Oregon along the coast. Also visited them many times out of body. They are in Hawaii now on the Southeast coast. Go visit them.

A few years back the elements told me I would go to a Grand Council being held at Stonehenge on winter solstice of 2009, If I remember right, to plan for the shift. I went out of body and met with other Shamans out of body, among other beings, to plan for the earth's shift. I was given the assignment of the North Pacific and a few other places to work on.

Torg, good job making pipe out of coral. Give to Kahuna. Not their tradition. Ryxi, learn to blow the conch.

When I go out of body, I feel extra tired the next day. I was really tired and ached all day so I knew I had to have gone somewhere.

Where have I been going out of body to?
All over.... Alice Springs, Stonehenge and Prescott.
Do other people go out of body while sleeping?

Yes!

Why can't I remember?

You will soon wake up! People think they're having nightmares. Age of Torg. You tried to visit Pongust at Stonehenge; his shield was not open. You need to warn him so he can open his shield.

What happens?

Bounce, ouch, wake up trashed. Druids made shields. Blondes need passwords, please! You visited Prescott (orbits the earth) too. He likes you. Whales are in Hawaii. You can now go see them out of body. You now have an anchor, go to South point where you visited and snorkeled with the turtle while in Hawaii.

Dec. 8, 2011

We're back.

Torg and Ryxi, you rock! Shaman class great!

What names shall we give to our new Shamans?

Bors, seeker of starlight and Mishu Domni, Sees within.

What language is that?

Adamic

What language was Torg praying in last night at our Shamanism 101 class?

Speaking in Adamic language, will happen more. Ryxi will soon speak Adamic too. Torg said, "I am here, send me" in Adamic.

Isn't that what Jesus said? Could anyone else have done the job Jesus did?

He was the best choice for the job. You too. Plan B not good.

What we imagine we will be doing is probably nothing like what we will actually be doing...

True

December 11, 2011

Mom hadn't shown up for our sessions in a while and I was getting concerned. She wants me to talk in the trees with Her more than have sessions with Torg so I can become more independent. I feel if we are both getting an answer together, the information is more accurate. She was also getting irritated with me because I sometimes doubt information. I wondered if She was staying away on purpose.

Elements: *Team here.*

Where's Mom?

You talk to Her in the trees.

Isn't it hard in the Winter? Aren't they sleeping?

Can still get messages.

I want a visitation.

She knows. It is rare even for prophets.

I still want to see Her if I am writing a book about Her to give me more confirmation I'm really talking to Her..

She knows. You have to be an open book and believe.

I think I do that.

Millions of people pray and get answers without God in person.

I know and many have seen Him.

Only less than 100 in all of time.

If I am supposed to make a big difference with this book, shouldn't I be able to see Her to be sure I'm really talking to Her?

Do you believe you have enough faith to see Her in person?

Unless I'm very naive, I think so.

Try to go on the Way. We can't go on the Way. Takes spirit.

"The Way" is a term the Nemenhah use as a place to visit people from other realms. The Nemenhah encouraged all people to go on The Way. I had friends who went on The Way and met Jesus, Gandhi and Panatan (a prophet and high priestess from the Mentina archives). I've gone there too and met Jesus among others.

Do you give visions?

No, spirits and Gods. We have no ears or eyes. Speak mind to mind. You would call it telepathy. Your mind sees pictures. Not ours. You humans rely on the physical too much. *If you can't use the 5 senses for something, it is not real. We have no 5 senses, all ether waves.*

I feel your feelings? Do you have emotions?

Yes. Develop more!

When you are a scholar you will understand.

I need to be a scholar to know how your world works?

Well! You call it dark matter. Can't see it. We don't really know, except it works. We make the wind blow, rocks move, don't know how.

Maybe it's like when we move our arms and legs, we just think about picking up a cup and our hand reaches over and picks it up. We don't know exactly how it works.

Is Heavenly Mother watching us?

Always. Loves you.

Watches all Her kids?

Yes.

Father?

Uses agents, watches some.

People want to know that Father loves each person and knows each person intimately.

He does. Mom, too. Different ways. Mom is about emotions and empathy. He is about intellect. All the Heavenly Family can tap into others. The group mind is one. You know your kids?

Yes, but not all.

You know them more than anyone else. Do your kids always see you?

No.

But you still know your kids. So Father doesn't need a constant focus.

When we talk to Father in Heaven, He knows us more?

Yes, raises awareness.

Will Heavenly Mother ever talk to me again?

When you are ready. Go on the Way, be a scholar, and use the trees.

In a session a while back I had gotten angry with Her for not empathizing with me about a personal issue. I kind of threw a tantrum and left so I wouldn't lose my temper even more.

So is She mad I was upset at Her for not empathizing?

No, She wants you to walk, then run, and then fly. Learn to connect in the trees more.

You both went out of body last week. Torg went to Australia. Ryxi went to the Stonehenge Grand Council meeting.

I don't remember.

You have a block, because you are scared. Haven't fully accepted your role of Ryxi yet.

How can I remember?

Accept who you are, no more doubt. A human committee met. Most don't remember. You don't see evidence?

Kind of, like Isis?

Yes. You pray all the time? I trust you to figure it out.

December 18, 2011.

I had ideas spinning in my head of all the things I wanted to do this year like when to launch this book, having a big gathering for the Summer Solstice at Mesa Verde and much more. Again, only the elements came, no Heavenly Mother.

To the elements:

Am I on the right track? Am I getting divine inspiration or are they just my own ideas?

Plan! Be flexible. We've planned for millions of years. Some plans ended happy, some not.

Like what I predicted. You don't know anything.

Yes, We do. Always the plan. No plan, no success.

Shamanfest at Mesa Verde, how effective will it be?

Very.

Solstice is in the middle of the week, shall I plan before solstice or after?

Before. Shamans gather, worlds change.

Mom says rest.

Easy for Her to say, She has no bills to pay.

She does, pays for the universe! Everything that happens takes payment. Jesus paid price for sin. His plan was for all to accept. Most do not. Do you think He failed? Mom and Dad plan for kids to become Gods, but almost none do.

How many people have lived here?

20 billion.

How many will become Gods that create universes?

A million. Some from other planets. Many come from Earth. It takes opposition to be a God, Earth gives opposition. More evil here to challenge.

Maybe that's why They aren't helping.

Yes. That's Her plan to make you strong. You're not alone.

Winter solstice - pray, visit network, many ripples. Rest, prepare, study. You must be scholars.

What is your take on finances coming?

Hope so. Had to happen. They're scared.

Why?

You're calling is light (information) *in the dark.*

Not me...

All you light bringers, yes "me".

I need all of you. Grid is way up. Even lawsuits are won by drawing from the Grid.

All light workers draw from the Grid.

What would happen if there were too much light in the Grid?

More of a whimper when it went down. Missionaries, attorneys, politicians, teachers. All draw from the Grid.

Chapter 14
Gaia Yoga

Elements:

You and all your team are nearly ready.

Like who?

All you can get! The big picture. Torg needed to remember the dance.

A few years back Torg was at a Shamanic ceremony where the group was dancing and he was drumming. The drumming put him in a trance and he was taken up in a memory of a previous time when he was a dragon rather than on the Human God path.

He and the other dragons were dancing the Ley lines into the molten Gaia. This would have been some billions of years ago. He remembered the etheric music, and has a visual memory of the dance itself. He would love to reproduce both somehow.

Why did he need to remember the dance?

He will need to dance again next year. Few dragons are left. Harder this time. Harder now because Gaia is solid.

What is the purpose of the dance?

Gaia Yoga.

Why Gaia Yoga?

She's sick like you (humans) *could be. Ley lines and nodes.*

So if Torg works on me, and balances my Chakra and meridians, that would be the same thing?

Close.

How is it different?

You're not covered with humans. Must be done gently, firmly, and quickly with minimal damage. All the work up until now is preparation for the shift to go smoother.

Gaia is shifting now, the dance will speed up and set the new Grid. Needs to return to its original pattern. Many points won't change much.

What made the lines move?

The earth spinning, tides, magic. Song, dance, ceremony, you're doing it. Changes are needed to keep Gaia alive. Minerals and air need the movement. Mountains, smokers, volcanoes need movement too. You Ryxi scholar? No movement, no new air, no new minerals.

Because we've used them up?

Partly, movement cleanses.

Like a stream, if stagnant, it gets sick?

Partly. If there is no movement, all suffocate and starve, oceans wash away land until there's all water and no land. All dead. Life in ocean. No food, no minerals, no life.

What about life in the ocean?

No. No food, no minerals, no life

Mayans say every 2000 years?

Partly.

The Hopi and other Native American tribes prophesy that Gaia will get rid of us?

No, this is different. Geology, not sociology. Some bleed over.

Even if humans didn't make her sick, this would still happen?

Yes. The first dance was creation, 2nd dance to break up Pangaea and after this dance, Gaia will rest, then crystallize. This is the last dance before she crystallizes for 1000 years.

After the Millennium she will crystallize?

Yes. The 1st dance was 5 billion years ago.

Did Torg dance the other two dances?

1st, yes. Not much 2nd dance.

Why not?

He was in spirit form.

Is that why Torg separates his intelligence and spirit? Rider too?

Yes, we need more dragons.

Why are there less dragons now?

Some left for other places. Some are now human. Just gone.

Can anyone be a dragon?

No.

Only certain intelligences?

Yes. Not you, maybe Rider, Kelly no. Leroy maybe. Your niece, yes. (I have a niece who is 10 years old and happens to have red hair, dragon hair.)

Does she need us to train her or can she do it on her own?

On her own.

How is she doing?

Better than Torg.

Is she conscious of what she is doing?

Mostly.

So if we said "Wow, are you ready for the dragon dance!?", what would she say?

She would freak out. Trust us! Give her hugs and chi work. Her essence will mimic. All merging of info.

Like a download?

Good Ryxi.

Who is this?

Finn. (Healing dragon.)

Are You keeping an eye on her?

More than that.

Was she a dragon before?

Yes.

There are lots of 22 year olds in my life right now. Everywhere I turn they show up! They are Dragons in Chinese astrology.

Yes, dragon energy. Time for dragons. You stop now and go to sleep. You will journey tomorrow.

How do you know?

Aurora.

Where are we going?

Not know. The fastest part of the shift will happen at solstice next year in critical places. Pray for your niece. Needs resonance reset at times. (hugs)

January 8, 2012 (1st time this year.)

Pongust is a very powerful earth element who has been guarding Stonehenge for a very long time. He and Torg have a strong affinity for each other. In fact he has been one of his guides. He often comes into our sessions.

Very good year, Friends.

Pongust - sorry, I was busy at Solstice. People are rebuilding Treehenge.

Have they built it?

No money to buy land. Some people don't want it. Think its evil. Resting. Deep breath. .

You go see Pichi? (Machu Picchu)

Need money. Highest priority?

Passport for Torg.

January 12, 2012

Elements: *How many good witches to change a light bulb? None, they are already light.*

You don't worry. Your needs are met, maybe not wants. We think you do well. Jesus did his whole work in a tiny place. Jesus trusted God. You will have your needs met.

Can Jesus make it snow in summer?

Yes. He makes his own flow. Every action and manifest has a flow and consequence. No action takes place in a vacuum. You bring Pahia, she's not home when others need her. The call went out.

Mystery school?

Doing it. Get more excited! Laugh, study, have fun.

Did I go out of body yesterday?

Yes, to see Pongust. Torg wasn't here so you went. Pongust does not want Treehenge built again! Old, bad energy. Soon, but not now.

What old EVIL?

Pongust: *Bad witches made babies and killed them. Good witches/shamans destroyed the bad witches and the henge. Worse than the Safe Haven one you know.*

Why did they do it?

Pongust: *Just power, Magic. Posts shaped like penises. Sex and sacrifices in both places. Gaia shook, was so mad, broke my circle.* (Stonehenge). *Now you know! Next time no* posts; plant power trees. None are over Treehenge now. Cast out. Water, semen. Rebuild in two years. Makes sex seem so evil. Powerful, not evil. Anything that can make a God can make a demon.

Was Treehenge always evil?

Pongust: *Original, no. Carved like maypoles.*

What made it originally good?

It was helping the Grid.

Have I gone more places this past week? I have felt exhausted all week!

Yes, opened a rift to another universe, brought back helpers from there.

You mean entities from this other universe?

Yes, like an Amoeba.

Amoeba?

Swallows you. Don't know from where. We couldn't follow your thread out. Thought maybe you not come back. Some Shamans not come back.

You were sent back with lots of riders, Amoebas. God is now assigning them all over. Happens lots. This is the first time for this universe.

What do you mean God is assigning them?

Like a student exchange program. They learn from us, we learn from them.

What assignments do they get?

Usually to watch, guard and protect. Some are assigned here to protect you and Torg, and more.

January 25, 2011

Elements: *Pahia is a big pebble too. Love amoebas protect you. Visit Mom, salt and amoeba work for shield. Joke: 2 elements went into a bar and the bar became heavier.*

Why are we here, to suffer?

No struggle, no muscles.

What about where Amoebas come from, is there struggle there?

No, NOT GODS. Amoebas don't advance.

157

Will they try to advance now?

Yea, they go back and create new. Students created them.

Do Satan's followers go to this kind of universe?

No, they go back to the void. Maybe new choices. No universes of light.

February 8, 2012

In the trees:

Pray to gain. Pahia is happy you are her friend. She loves you. We love you, too. Why do you question? We want you to be happy. She is called to help/work with you. When she comes, the earth will quake. Why you worry?

I feel I'm not worthy.

Nonsense! We love you, you are called by Us to do this work.

Anything else?

We call you Pahia. Look it up.(I remembered Pahia said her name meant "Happy".)

Was I reincarnated from Hawaii?

No, a guide, sisters. What you do makes the world go round.

Not literally!

Yes. Know yourself!

Thank her. Write her a letter and mention all the things you remember .

Thank You for Your love and support.

Welcome. LOVE YOU! You deserve so much!

How am I doing with anger?

Better, lots to go, but better.

For the past couple of months while walking through the trees, I asked when I should bring Pahia here. I got February and that she would come to fulfill prophecy.

With much work, our plan to bring Kahuna Pahia to Utah happened. I posted on Facebook asking for help with air miles to bring her here and someone answered the call. We spent much time on the phone before she came trying to get all the details confirmed,

158

but I still felt there were a lot of loose ends. As much fun and good her trip brought, it was also not a bed of roses!!!

The upside of her visit was that she brought teachings, blessings and awareness to Utah. She was a guest teacher in our Shaman 101 course. she was able to travel to Arizona and New Mexico to fulfill a prophecy of a Kahuna meeting with the local tribes sharing teachings and prophecies.

I scheduled many speaking opportunities where people came and were very excited to meet a real "Kahuna". It was extra special to me that she was a woman and an example of a woman leading. She scheduled blessings and people loved them. Most of her teachings were to share the "Aloha Spirit", that "sharing is caring and caring is sharing." She brought many gifts for people, like chocolate Macadamia nuts, etc. She always wanted to make sure people who helped got a gift. She brought beautiful things for people to buy at the events. I had made the suggestion based on how a traveling Shaman from Peru created funds for his travels. I didn't want her to go home empty handed. We also wanted to have her be a co-teacher for Shamanism 101 and we would split the tuition fees. Our half would go to funding our travels to nodes, Shaman world tour.

We had many students who took our Shamanism 101 class because she was a guest teacher. She explained how the Hula dance was sacred and not meant for entertainment. She shared a dance with us. Two other Hawaiians took the class and danced with her. At the end she asked what else we came to Hawaii for, and I told her to get a name for me. She then said again," Whale Rider" and that we would need to do more of an official ceremony to make it anchored.

Unfortunately, whenever there is a lot of good, the challenges appear, too. Her paranoia of whites taking advantage of Hawaiians, "Hawaiian time" vs "western scheduled time," her being offended over innocent mistakes, my worries over "doing it right" and our not having an agreement in writing before to be more clear about expectations all created a lot of tension for us.

Because of her ancestral experiences of whites taking advantage, she was ultra paranoid of being taken advantage of by us and many who showed up to help. One of her friends had been talking to her before she came and had planted seeds of fear in her mind. When she got off the plane her demeanor was totally different from the Pahia we met in Hawaii. At one point she even accused us of bringing her here to make us look better.

I had never hosted a Shaman before and we didn't have time to write up a contract to specify details so there were a lot of misunderstandings. It was also stressful being the bridge between our culture of tight scheduling our busy days and her Hawaiian culture of going with the flow. When I scheduled events our people wanted to make sure we started and ended on time. The Hawaiian culture and Shamans in general go with the flow. That left me holding the bag of trying to keep Pahia on time, and keep our audiences from being upset that we weren't starting on time or ending on time. I was so stressed by the drama that by the time it was over I never wanted to host another "Shaman" again!

It didn't help that Heavenly Mother was still only talking to me in the trees. I wasn't feeling very confident in the accuracy of the answers I was getting.

Feb. 20, 2012

In the trees.

Hi, Ryxi, I LOVE YOU. How are you?

Happy birthday. Proud of you. Can we see your progress?

Of course! Aren't You watching?

Yes. We need permission.

That sounds odd.

To go into your mind.

Oh. Sure. (Stayed for a minute.) This seems weird, are You sure it is You?

Quite. I'm wearing red shoes, green tie.

Why red shoes?

Because I want to.

Am I on track?

Yes, always.

March

I was still being left with only talking to mom in the trees. The trees were still asleep for the winter so they weren't super clear

160

either. Even Torg was not doing sessions with me, with the elements and our guides. He said he wanted me to develop my skills, and faith in my skills, by myself. It was frustrating!

One day one of our earth Grid team members I hadn't talked to in a while called. The elements call her Elf cause she has elf ears and they say she has elf ancestors... weird. The elements talk to her a lot, and Pongust especially will pop in and visit with her.

She asked," Have you spoken to Pongust lately?". I told her I was being challenged to connect on my own and hadn't connected lately. I guess he was irritated with Torg for not working with me because he told her he was acting like a baby! Ha!

She was excited to hear that I was trying to learn new ways of connecting and she said she was too, and wanted to share what she was up to. She had been watching a video called *Abraham,* by a woman named Ester Hicks who channels a person calling himself Abraham. She wanted me to watch the video. I watched it and found it interesting how she channels. It gave me more ideas on how I can connect.

It was obvious I was in the flow of the Universe to learn to connect, because our friend Hunter who is a channeler, also called me and said Mother Mary wanted to channel through me. She was channeling through him but wanted a female to channel through. I was honored and open to the possibility.

He came to my home at 3am with Kelly and Rider. He brought a picture of Mother Mary to help connect. We meditated for an hour and I felt very strong tingling sensations but didn't channel. He thought it would take practice.

Then Elf called again to tell me that someone named Seth had been visiting her. I checked in and got that he was a grandson of Seth in the Bible. He connects with her through pen and paper. He guides her hand to write messages.

She informed me he wanted to connect with me too. She told me to get out a paper and pen. And start meditating. He was telling her on the other end through writing what to say to me.

After a few minutes of meditation my upper body started circling like it does when I'm facilitating a sweat lodge. People in the lodge don't see because it is dark. I'm glad because it might look weird.

161

She said," He says you are circling and almost ready." Wow, how did he know! "Now is the time" she said. So I picked up my pen and put the ink to the paper. At first I got circles. She said it would take a while and that she would hang up while I wrote to free my hands. The pen started making an infinity sign, then a spiral in and out. This went on for an hour. Then I was told to go to sleep. That was enough for now.

A few days later Elf came by so we could work together. Hunter just happened to stop by at the same time! No accident. So we decided to have a session together.

Dragons showed up and told Hunter and Elf to be at the next sweat lodge; they had a message for them.

April 3, 2012

I scheduled a two sweat lodges in a row because the at last sweat we had, there were so many people, we couldn't fit everyone in. Hunter and Elf showed up to connect with the Dragons. One of our Shaman team members, Maire, ran the first one. It was nice to let another person run the lodge so I could have a chance to see visions.

I got a vision of the earth opening up showing records. The records were downloaded into my mind and body. Then when I was leading the next lodge, Maire sang the dolphin Chumash song of ancestors, which is about whales and dolphins being the ancestors. As she sang I felt myself riding the waves and saw my whale. When I spoke of my whale being there, Maire reminded me they are also record keepers.

Then I remembered the elements told me my new Hawaiian name is "Whale Rider". Kahuna Pahia also got that my name was "Whale Rider".

Then Maire had a vision of petroglyphs in 3D floating and saw them in someone's mind (records). I turned to Elf without knowing why, I said, "Are you translating them?" Her eyes went wide in surprise that I knew and she nodded.

April 7, 2012
(While at church)

I like your necklace (Trinity triangle w/spirals) Come see me tonight. Much to tell you, Love You. You are getting used to being in my presence.

Ryxi Chick the best!

Be more angry about souls that are lost. Not angry, but concerned.

Have fun, I am with you. Be ye of good report for I will comfort you and strengthen you when you are in need.

That sounds kind of cliche'.

Yes, but true. Make your life work with me in tandem. You are on the right track. Keep it up! I am sending many to you, and I will give you the words to speak and the blessings to get it done! Be still my little one. You are my child and I offer you eternal life.

Still sounds cliche'.

Yes, but true.

One of my friends had a visitor named Seth come to her. Who is Seth?

My son who was brought here to challenge my kingdom for good. But now is minding his own business and calling all to repentance, that he might join me in eternal life.

What are all these tingles?

Energy sparks that ignite when they merge with like sparks that come from us. You are healing and growing so fast. Just in Time! Good job!

April 7, 2012

Torg informed me that it was now time to work together again. We hadn't worked together for a few months. When we opened the session with prayer the first person who showed up was MOM! She was back! I had sooo many questions!! I wanted to confirm all the information I had received in the trees. It had been four months since Torg and I had talked to Her together! For the next few months she would come into our sessions frequently.

Pahia.

Her personality. Not typical for all shamans.

163

I hope I got a lot of God dollars for that. It was very stressful.

You did. You're on a new level now. WOW!

You much bigger now, keep grounded. I'm proud. LOVE YOU! You are doing great now.

Is it OK if I talk to You on Wednesday?

Yes, not a question. Silly kids. You talk to the trees first, then ask me. Like today. Good girl, no spank!

Torg?

Good call to listen to music. You grow, too.

I feel like I am back in Kindergarten.

You are; second grade!

What triggered this change in me?

You grow quick. Very proud. Most people experience the mighty change, and freak out and quit.

I saw a Facebook post that helped. It read: If you are depressed, you're living in the past. If you are anxious, you're living in the future. If you are at peace, you're living in the now.

Your change helped Torg to change. Torg now understands music. Say, "Thanks Ryxi".

What made me grow?

Pahia, mom, Hawaiian Sunlight. Growing and accepting your role as a Shaman. Your mom is better and accepting your calling more.

Proud of you for embracing.

April 10, 2012

In the trees:

Is channeling OK?

Yes, but risky. Others can get in that are not trustworthy. Not recommended. Not wrong, just not best. My kingdom has many possibilities. Not always set in stone. Like you, always changing. General guidelines and basic rules. Sometimes rules are broken and turn out for the best. Like you. You make things happen by bending rules sometimes. We watch over you. As long as your heart is with Us and in good intent, We are happy. Make the most while you are here. We need you here and you can soar easier later. So happy for you and

*your new life of abundance. Keep it up! Many new
opportunities coming. BLOW YOUR MIND! Go talk to guy
with a limp in sanpete with records found in the caves.*

Later with Torg:
Where have you been?
You needed to learn new skills. Torg might leave soon.
**In the trees you mentioned someone, did you mean the
guy who found all the artifacts in Sanpete county?**
*Yes, he has lots to tell and show you! Maybe tomorrow?
Soon!*
Any feedback on Pahia's visit?
*Pahia has issues and a calling. Yes, need boundaries.
Learn from her. Always be humble, teach and learn.*
Did You send her or did I just want her?
We sent her.
What do you think of our friends channeling?
*Fuzzy, not unrighteous. No true prophet gives
permission for possession of spirits. Rules provide a
foundation.*

April 14, 2012
*I'm wearing blue with green hair, red shoes. My kingdom
has many colors besides white.*
What does the infinity sign mean?
Means all for one and one for all. Never ending. Eternity.
*She sang and danced to: Come follow me, my Mother
said, and in the womb my children spread all over earth and
skies above. In my own time I will bring love.*
*You and Torg need to find a sick tree in the park. It's
affecting the park and spirit. You must fix. Affecting
messages from Heavenly Mother.*
My nails sparkle!
Does Satan have a wife?.
No mates in the spirit world. Marriage needs a body.
**What about married people who die after here and are
spirits**.
Friends, not mates.

165

We are getting ready for Summer Solstice at Mesa Verde, I wanted to have a festival called Shamanfest and bring Shamans to teach students and share with each other. Maybe invite Kucho, the Shaman from Peru to join us.

You fix the Grid. What should be, will be. Only 20 shamans needed. Better to have a good team of a few than lots. I call many but few show up. Too many would make a circus of a sacred event.

What do You want me to do?

Teach ceremony. You do what you want. All I need is a few people for the ceremony. If some are not on board, it can lessen the effect or make it bad. Can you and Torg keep people in control? This is your Grid spot. Do you trust all the people that will come to your festival?

The solstice is in the middle of the week, do You need the ceremony exactly on that day to be effective?

Wednesday is best, but Sunday is OK. Best to have good people on board on the same page. You lead, no conflict. A big Festival might make the Grid weak. May be evil people there to tear down purpose.

Aztec dancers at solstice yes. Ceremony no charge and yes, charge for classes.

You always encourage me to have confidence. I want to make sure I don't cross the line between confidence and ego. How do I tell?

Ask yourself: "Do I care if anyone knows it's me". Ask what Jesus would do.

Is it true that we don't need to sleep?

Heaven's No! Where did you get that idea?

A friend.

Maybe when you become a God, not while Human. Your friend is confused.

Hunter channels a spirit named Jahia. Who is Jahia?

My messenger of truth.

A Holy Ghost?

Kind of. For certain people, yes. Special for Hunter.

Did he help create the earth?

166

Yes, among others. You too.

All of us?

Some more than others. You did a lot!!!

Pahia?

Healing. She is trying to tell people what happened to the Hawaiian's and change the world. Hopes to make a difference.

What is the prophecy?

Truth.

Yeah, what truth?

You'll see.

No hints?

Guess.

Jesus visited other places. A little about where humans came from?

Good Ryxi. Mostly right.

Who are the pleiadians?

A planet in outer space.

In this universe?

No.

April 2012

Tomorrow, find the dark tree. Praise the tree, it has had a hard life.

What happened?

You will see.

Use a crystal?

Try. Good idea. Burn the wood in your sweat lodge to honor it. Give the deva to the Paulownia tree.

Wow you guys! I'm wearing a white robe going to a meeting.

So when You are with the Council You are conservative?

You got it.

We went for a walk in the park and found the tree Mom was talking about. The tree was growing in all different directions. It actually had a unique style of it's own, but we could see it had been abused.

A few years back we learned a tree spirit, a deva, can jump from a tree to a crystal and then be transported to a new tree.

We sang to the tree and honored it. The deva transferred into the crystal and we placed it next to our fairly new tree in my front yard. It was so happy now. The tree grew more that year than ever before!

April 23, 2012
Do you know why I am irritable?
Space invaded. Trigger. Hormones are a bomb.

As always there were a lot of people hanging around my house between renters and friends visiting. Torg rents the basement and has many clients for body work coming by. My house is like a community center, mystery school. Usually I love having people around. I grew up the eldest of eight. But recently I was irritated and just wanted to be alone.

I'm too compassionate.
Not. Still need boundaries.
My hair is rainbow.
Torg's dad (who is dead) visited Torg and says there are only 9 dimensions there.
That he knows of. Not that many more. Thousands in other universes. This is my generation.
Hunter's friend says you split yourself off.
I can appear in many ways. I can shape shift. No hologram. You can, too. Mother, sister, daughter, cheerleader; you look different as a Sunday School teacher than as sweat lodge facilitator, etc. You can learn to project the reality you want.
I can make myself look like a hot young lady!
You are! (Hot) *I do the same, but much better, of course.*
Ego?
Only truth.
What's the difference between ego, pride, etc?

Depends on the purpose. Made you laugh. My gown is low in the back. I like Father's hands on my back.

How is the Deva?

Good job. The park is clean. Now is in front yard and healthy.

Cake is fun. Chocolate with cherries.

Anything else?

Whipped custard. Elixir dribbled chocolate.

Pomegranates.

Do they represent You?

Yes, fertility. Put in the sink at Mesa Verde with Gaia. Help you communicate. Chocolate, pomegranate, cinnamon. WOW.

How do You know? (Before, She had not tried it.)

Because I am God, I know. Are you questioning Me?

April 24, 2012

Ark of the Covenant? What is it?

A box. Like in Indiana Jones.

Where is the Ark of the Covenant?

Won't say. Won't say. Safe Place. In a cave. Later it will be found and put in museum with other artifacts. Includes Jesus' artifacts.

What is it?

Many meanings. Read the Bible. Some say Mind with God.

Why all the wonder?

Some gold & silver and precious things. Scrolls, tablets, crystals, herbs, ashes, cell phone. A transmitter to God. Includes Jesus' artifacts.

Transmitter like earth portals?

Yes, Stronger. The hand of God killed the poor guy, oops. Made the king study.

Poor guy died, but people were saved?

You study it.

Why carry it everywhere?

A Sacrifice to remind them of the covenant.

169

Includes Jesus' artifacts? I thought the Jews didn't believe in Jesus?

YES, Jesus artifacts. Was opened since Jesus.

How did they get them in there?

Opened it!

I thought if people touched it they would die?

Jesus opened it.

What are the artifacts?

Writings of the Millennium.

His writings?

Yes.

COOL! Will they come out at some point?

Yes.

It would be so cool to find and open.

It will kill you!

Why do we need it if He will be returning soon?

Physical witness of the historic Christ with the modern Christ. Jews will open it.

Why will it kill me if I open it and not kill the Jews?

They are chosen. I should have said a Cohen Levite.

Before Jesus comes back?

No.

They open and know for sure?

Close. They need to read it. He told of his own return.

So after Jesus returns they will open the box and He will have told of His return, and prove he is the Christ?

Close. Last sacrifice, heifer. He will also have the marks in His hands and feet as proof.

When Jesus came to me in a vision, He showed me His bloody feet. Was that a real vision?

Yes; because you are blonde and for you to know it was Him. Ha, ha.

Well, there must be something good about being blonde.

I AM.

You are too funny.

IAM.

What about the Holy Grail? Some say it is the Last Supper cup?

What cup? No such thing, not like people think.

What did they do with the cup?

It was washed and put back in the cupboard. The Holy Grail is the blood line of Christ. Silly people looking for relics!

I want to make a "thank you" book for my Mom who is in the hospital but I don't have time between Your book and everything else I'm doing.

All efforts are worth it.

Who's the priority?

Your Mom , your mother-in-law, then Me. I've waited billions of years. Another month is no problem.

April 25, 2012

I've been asked to speak about repentance at my church on Sunday, will You give me inspiration?

Of course. You need to learn to speak. A good subject for you.

Because I need to repent?

Not much.

What do You want me to talk about?

I like what you have so far, it's what the people need. The worst sin people commit is to be like Satan and force others back into our kingdom. I will help you tomorrow.

Money coming?

Yes, 2 weeks. (Money came in two weeks)

April 30, 2012

Pongust - *Getting ready for the shift. December 22nd is the peak. More, more, more . . . Peak . . . less, less. Most of the peak will be within a week or so. You plan your stuff on the 21st of December.*

Where is the best place for us to be?

Here, to protect. You've spent years making it safe here. The Winter solstice is very dangerous.

Not sure anymore what dangerous means.

Torg could die. This mission is the most dangerous of all time. If he is not protected, he won't come back.

Should I dance too? I went to Madagascar last time even though I was supposed to stay.

It's best you protect them when they go. Remember the last time? Wasn't the best idea. You were almost lost, Ryxi. Almost didn't come back!

He's talking about when we took over the node in Madagascar and turned it from dark to light. I was supposed to stay and protect with women in the lodge, but ended up going out of body with him to give more protection.

Am I protecting their bodies?

Their Soul. A spirit without a guide dies. If Torg's intelligence doesn't come back, his spirit and body will die. He needs a spirit and a body to come back to. If the way back is unsafe and the body and spirit aren't protected, he will stay a dragon and never come back. He won't be a God. He'll be lonely.

Will his intelligence go back to the void?

No.

Why would he be lonely, aren't there other dragons to hang out with?

Yes. He would have no mate, no kids, no other universe, etc. Can you do this? His life and death depends on you!

Of course, how?

Six men, six women in a circle. No break until dragons come home. (See Prescott clarification November 28.) Dragons dance for the good side and dragons dance for the bad side. We must win.

If we don't, what happens?

The plan fails.

How?

No new Gaia. We've always won.

Casualties?

Need everyone. Don't break the circle until the dragons are done. Pee first.

How long will it take?

Moment or eternity.

Who do You want for the circle?

Mostly humans. 2 elements.

Who?

Me (Pongust) and Nolmi (Great Salt Lake Salt element).

How do we hold your hands in the circle?

Show you later.

Won't Stonehenge be vulnerable?

Yes, big trust I have.

Mom: *Mom here now.*

Who was the Polynesian I saw in a vision in the sweat lodge?

From Tahiti, wanted to make a connection, a physical person.

You need to know as many as possible. He is a Grid Shaman. All Shamans heal both earth and people. No box, no narrow view. No numbers.

What will happen when the dragons dance?

Don't know. Lots of possibilities.

Will we release a big dragon?

Ha, ha! Might! Just do your job!

How was my speech at church?

Proud. You were nervous. Told people what they needed to hear.

What do I need to practice?

Nervousness? Flow of words.

Why should I make plans if You keep changing them?

You learned by planning. You are better for it.

Learn all things. Keep adding. You become a scholar, you get ready, you're not finished. How do you answer questions if you don't know the answers?

May 2, 2012

I'm in a better mood.

Me, too.

Like Mother, like daughter?

Yes. Fruit falls close to the tree.

Nice saying.

Godly.

I'd like to translate the Bible, etc. Clarify.

Nice hair – Goddess.

Do you watch us all the time?

Not in the shower. Not while you're picking your nose.

The new millennium is about family, relationships, and new information. Everyone a scholar.

Isn't everyone able to be a scholar?

Not yet. Some can't find a way to feed hunger for knowledge. Have a desire, but no access.

People need internet access.

More info is not enough. Must have wisdom to use the info. Imagine a planet full of Crystal children and above.

What about those who create? Always Crystal or above?

Always. You are. You're not at your potential yet but climbing. You were very close weeks ago. Many kinds of Crystals. You were born early.

Born early to bring in Millennium? Is Torg a Crystal?

Sort of. A Dragon is a special case. Head case.

Your description?

Search and can't rest, need projects. Need to make a difference.

That's Torg.

Tru-ish. Many advanced before. Crystal is a term for the group arriving now. Large numbers now. Most dragon who chose to be human are Shamans here now.

What skills do dragons have?

Shamans are mostly crystal children/people.

Are all dragons advanced like crystal children?

Not all. Some lost their way. Not all crystals kids are dragons. Dragons all through time danced. Dragons born now are Crystal dragons. Dragons are not Crystal kids unless born in the last ten years. Close to the New Age description.

Reading minds is a sign of an advanced being. Advanced beings usually want to make a difference.

Torg: Before Crystal and Indigo, were "creche" and "renaissance" Your terms?

Bingo.

After Crystals? Maybe Platinum? What's Your term for it?

Special kids. You were born in creche time.

Ryxi: So I'm a Crystal type born during the Creche time with Creche and Indigo traits, too?

Creche are adults when the shift happens. Most Indigo's are too young.

Are most Crystal kids 22 years old? (The kids that keep coming around)

Very funny. The Crystals will help rebuild and send knowledge.

Me?

No, you bring shift through safe.

I thought I was supposed to build and bring knowledge too with the wisdom centers and Mystery schools?

You will much overlap. Your grand kids will be the next level.

How many levels are there?

Gifts are many. You wouldn't even believe it!

Will most of these advanced kids coming be the Gods creating universes in the future?

Not really, some are really good, some are really evil.

Interesting, the New Age community thinks all Crystal kids are Good.

What did you mean when you wanted me to be more concerned about lost souls?

None are resurrected in time to help with the change into My Millennium. In My Millennium We will search and find them and clarify mysteries. Information saves. More will know about commitment to Us.

So less souls will be lost during this time because they have more info?

Yes, you are right

.What is your definition of a lost soul?

Not resurrected, confused.

So, lost spirits who haven't gone to the light?

Yes. Also spirit prison. Jesus went to the lost souls there and many followed Him. Noah's flood filled spirit prison. Not all wicked, but all lost for now. They are in time out!

May 4, 2012

I was having a lot of challenges so went for a walk in the trees.

Holy Moley. Meet My Son in heaven when you finish your work. He will consecrate you His. He is more concerned about your mission than His love for you. He will stand back and allow you to fail if it needs to be. He loves you and would love to save you from failure, but wants you to grow and be all you can be. So He's willing to endure the pain of watching you suffer and fail, then rescue you, than have you not grow.

Can't He help?

No, not all the time. Encourage, yes. Test, no. It is a test. Otherwise you would not be strong enough to be a God and stand on your own two feet.

Don't You ever carry us?

Only when you do your all first. In balance.

May 6, 2012

Welcome home!

I had to ask a tenant to leave, I feel bad but she is stressful to be around. Any thoughts?

No, I drowned mine. I burned a few, too. Tenants are such a pain. And kids.

Tomorrow I will give a speech about You for Mother's Day!

Proud. Wow! Take grapefruit oil.

Why?

Calm and focused to listen to me. Put on forehead, third eye.

Why do I want to be a hermit lately?

A normal phase. You are growing. Won't last long.

You kids don't listen to Us; like Moses with the Ten Suggestions.

I haven't been studying all the subjects You told me to study; like chemistry etc. I don't understand why I need to be a scholar. Wouldn't it be more believable that I am really talking to You if I am not a scholar?

I have an idea. Maybe you could have a panel of experts to back you up.

I like that idea.

May 8, 2012

I spoke at a local spiritual gathering about Heavenly Mother for Mother's Day and they LOVED it! She obviously did too.

WOW, WOW, WOW, WOW, WOW. High 5!

What was the highlight?

Not afraid, stood in power, speak truth. Poor Torg; all those girls! Worked. He felt the drum from Me.(Torg drummed and was especially inspired)

Your cookies will be done in 2 minutes. (I had cookies in the oven)

Now you're a cook?

Not cook, kook!

You made me sound like a stand up comic!

You are!

I shared with the crowd my favorite part about Her...how funny She is. They loved it!

Last night yes, yes, yes! I like it yes. We each gave you a score; 9, 10, 10. Dancing with the Gods. (She likes dancing with the stars).

Who gave me a 9?

Dad. Stiff Brit.

What does he think I need to work on?

More polish.

Jesus was the 3rd judge? Was he there?

Yes, He baptized you and keeps an eye on you when I tell Him to.

177

What do you say to women who are in abusive relationships?

Leave.

If we had women leaving abusive men, they would straighten up.

Yes, both ways. Consequences. Relationships should nurture. Both ways - parent, child, friends, nations.

Yeah, me! Did I ever tell you I was a cheerleader? I'm better looking now, you should see me in tights!

What could I do better?

Don't lose your glasses. Just more finesse. Maybe Dad will give you a 10. Go hug your mom for Me.

I'm in tights now. I go see Dad.

Fishnet?

Not yet. . . . Now they are! Bye!

I didn't get to work on the book for Mother's day; I had a Million things to do.

Just 999,999 things.

You spanked your tenant just right, no anger. Yeah, Ryxi! You get pat on butt. Torg do it for me?

I'm happy about your idea for the panel. Much easier now.

You studied a lot. Now a scholar about me.

All speakers be a panel for each other.

I think, therefore I AM! I am Goddess, hear me roar!

May 10, 2012

MOM - feel better?

Yes, what is the stone Torg found while digging the basement?

Like a djinn/genie.

What is a genie?

Close to a phoenix. (medium sized fire element)

It was in the rock and released when dug up. It has a large ego, likes to show power, envies dragons. Dangerous but not evil. Solomon bound many to keep peace in the kingdom.

They are not demons, Pongust is wrong about that.

Where are they from?

From the mountain. It was bound and buried. Many of them.

Pongust said a giant threw it out of the mountain and bound it. Who was the Giant?

Pax?

There are three Giants (powerful earth elements) in the United States that we know of. One guards the four corners area (named Julie), and one guards the plains (named Pax). Pax is the peace maker and Julie has "PMS" a lot. He keeps her balanced. She has had many evil wizards and sorcerers try to control her and is a force to be reckoned with! We met her on our travels to Hopi land. She loves the Hopi because they dance for the earth and keep the Ley lines charged. Gaia calls the Hopi her kids. Once Julie told me to bless our Hopi friends by baking a cake in their oven.

Where is it now?

Free and safe (to have around). *Ready to behave and act like a phoenix, not try to fly like a dragon. No more time out. Can cause or put out fires. Causes good or bad dreams. Not trustworthy for wishes. Be careful what you wish for.*

The rock?

Move it away from plastic and computer. Move to dirt.

Where is the genie?

400 feet Northwest In a garden. It was told to go help the garden grow. Be a friend and give it chocolate to ground it, and mint to make it strong. You may need it, too. It owes you.

What is it helping grow?

Berries.

What will it help us with?

Growing plants. Genies don't fly, and are frustrated.

Aren't they like phoenixes?

Phoenixes fly, burn out fast, fly again.

How are genies different from phoenixes?

Not mature. Do magic to impress.

I think the reason people think we can only pray to Father in Heaven is because Jesus taught in the Lord's Prayer to pray to Father in Heaven and left out Mother in Heaven. Wouldn't it be more balanced if He would have said "Mother and Father who art in Heaven..."?

Not for the Israelites! It's more complicated.

Seems out of balance.

Was!

Elohim is plural for God, why not use Elohim?

It was illegal at the time to say the name of God. Could be stoned. People would have flogged or killed Jesus. The new church would have been dead in the water. They did well with what they had.

Was it ego?

Not about ego. You have more information now. Do as well as they because you have more now.

Did they know about You?

No, not much. In their day Goddess worship was extra evil. Prostitutes. Satan did a good job of twisting. It's complicated.

Yes, you slowly change. You're on your own and fly. Fly like an eagle. You have a new role soon. More smiling, singing, etc. Break through. Forgive all!! Many come to you for advice. Make sure you know what you are truly about.

Why can't I just ask You?

Yes, but you need to learn treasures for yourself.

Learn what?!! There is infinite knowledge.

You will be asked to exercise the keys of Enoch.

Which ones?

All!! Flying.

I didn't know he did that.

Lots you don't know.

No duh!

Many will choose to stay. Many will choose to go. They will ask you for directions. Make sure you wear your seat belt.

This seems outrageous!

It is!

May 16, 2012

I always like to test my answers. I don't know why, I should know by now this is real. It was my friend Rider's birthday and I wanted to get something unique for her. So I went to visit the trees in the park.

What shall I give Rider for her birthday? I want something unique that symbolizes gratitude.
  A potted flower with meaning.
Suggestions?
  Lily.
What's the meaning of a Lily?
  Gratitude.
  Teach people how to talk to me. Give them pads of paper during your workshops and try to be in the trees.

I love it when I get confirmation!!! I went to the store and found two lilies. A purple one and a white one. I chose the purple. When I gave it to Rider she said, "How did you know? Lilies are my favorite flower! I was just looking at those at the store and wanted that exact purple one but decided I shouldn't spend the money!" I was thrilled I was getting the right message, how could it get any clearer than that? Even with this proof I would still wonder...

May 20, 2012

In the trees:
  Have fun while you live, even in times of peril. Make no mistake, much is yet to be seen. Much more information is coming. Be faithful, my faithful servant. Be patient and live moment to moment. I am with you. A clouded head makes for clouded judgment and desperate actions. Play, sing, dance, celebrate. For I am with you, and cradle you. Much is yet to be seen . . . make sure you know where you are going and where you came from so I can be your reward sail. Every ripple, time will tell. Let me take your stress and don't let it happen again! My kingdom on earth as it is in heaven.
  (Visual of a group sitting around a table restructuring and negotiating.)

181

A business I had invested in was struggling and I was sad because I wanted to use the money for Australia.

> *LOVE YOU RYXI! Hang in there. I won't abandon you now, my faithful servant. Keep chin up. Tomorrow is a new day.*

Australia?

> *Right on target. No worries about money; coming. Make sure you follow your heart and stay in alignment with Me. I will make sure you get to Australia.*

We need $3,000 to go to Australia.

> *On it.*

Later that day $3,000 showed up.

May 21, 2012

> *I was right about money.*

Yes, Thanks! Any thoughts on the weekend?

> *I Love the dragon shed!* (A sculpture I designed on the outside of an earth building at Safe Haven inspired by Jesus, the white dragon.)

Did You have a good Mother's Day?

> *Mostly.*

How could it be better?

> *More of you praying and listening. All moms want kids to listen. You want your kids to listen and your mom wants you to listen.*

> *Use priesthood to get wisdom.*

> *Part of your priesthood is to pray as a mother with Me to get understanding. Each kid is different. So is each Mom.*

But if I listened to my Mom, my life would be way different, not what I'm supposed to be doing. Does that ever happen to you?

> *Yes, dammit.*

Do you pray to your Mom and is she always right?

> *No.*

> *My toe nails sparkle! My hair is chealla.*

182

Color?

Rainbow. Dad rolls his eyes.

He likes it, right? How crazy You are?

He'd better, I know where he's ticklish!

Torg, take the onyx home stone you picked up from Adam's temple and take it back home to the island. (Philippines)

One came from there. Stonehenge, too.

It's not formed in the Phillippines?

No, but belongs there. Was formed in the Middle East.

Makes sense, Enoch's formed . . .

No, Enoch's onyx was formed there. In general, onyx's are taken around by different people. Onyx makes anything good stronger.

Black only. Red, mostly. You remember the Enoch stone at Giza? Evil didn't try to find and use it. It only wanted to stop good from using it. Enoch's onyx is from Montana.

What about the ones on Israel High Priests' clothes?

From Asia. Torg's came from Australia. - Canada.

Probably to connect places.

Smart witch. Find in Matrix.

Pongust will be here soon, in a red rock.

When?

Won't say.

Will the courier come to the door?

No. Needs to be secret.

I would love for him to come to the door!

Watch what you ask for! Take Pongust to Eve's temple. You release him from the red rock when he gets here. He's sealed and protected so no one follows. He's very strong. Honors you much. He is a Djinni, like earth element. He could have been very bad if he had gone evil.

Pongust isn't one of our five honor guards, but is honored and trusted by us. I will tell you how to release him when he gets here. Write symbols first. 7 symbols and an enabler. A Druid viking is bringing him. Bigger and younger than Torg. Blonde.

Who sealed Pongust up?

A Solomon descendant.

Does this person know us?

Yes. You may meet him when you go to Stonehenge to bring Pongust back after he travels with you.

Let the Djinn help you.

He's now in Wasatch Garden. Comes here every night.

What does he do here?

Watch! Magic!

We were at Safe Haven for the weekend – A new person was there and sang an amazing song. Torg thought it was kind of like the dragon dance music. What did you think of Marie's song?

Wow, she's special.

Keep her around?

Try. She is like Torg, looking for her tribe. Trying to re-create the dragon music. Original dragon dance. She composes music. She has much to offer.

Rededicate Safe Haven every six months.

May 23, 2012

Pongust will be here in the morning. Let's talk then.

Will the courier knock on the door?

Doubt it. Up to him. His name is Mac, a Scottish druid.

I want to meet him! Should I leave him a note?

Yes.

Does he know who is in the package?

No. Knows it is special. Knows the sender is a priest. He makes many secret deliveries. Most in Europe.

Will the note freak him out?

No. He is used to it.

May 24, 2012

I was so excited that Pongust was coming and that someone, a human, was bringing him all the way from England! One more confirmation this was all real! I wanted so much to meet this "Mac",

184

so I put a note on the mail box and a note on the door saying,"Mac, please knock. Ryxi." I slept on the couch to make sure I heard him knock, or at least so I could see and hear if anyone came. I felt like a kid trying to catch Santa Clause!

When I felt the sun shining on my face through the window I knew it was morning. I sprang up and ran to the door to see if my note was still there and to see if there was any sign of Mac.

The note was still on the door and the note on the mail box had been been flipped over and shoved into the box! I opened the mail box and tucked inside the note was a 2" round bundle wrapped in brown paper and tied with a string.

I ran downstairs and woke Torg up. We opened the package together. It felt like Christmas! I unwrapped the brown paper and found the next layer wrapped in foil. I unwrapped the foil and found a red-orange kind of squarish rock just like Mom had said, it had symbols written in black magic marker all over it.

We almost threw the paper notices out, but Torg happened to glance at the one from the mailbox in the right light and saw some scratchings on the bottom, as if the courier had used a pen with no ink. The words were, "Seamus McCormick, hi!" OK, so Mac was the nickname for McCormick! Needless to say, we kept the paper after all!

We remembered that Mom had wanted us to talk to Her before we did anything else. So we said our prayer hoping She was available.

He's here, safe!
Who wants to carry him?
Torg is best, I lose things and he is more connected to Pongust; he's been his guide for the past few years.
OK, you first need to ground and open your root chakra.
How do we ground, stand on the ground without shoes?
Feet flat on the ground, concrete or on dirt is OK.
(Torg's basement room floor is concrete)
Don't laugh! Ryxi, spank Torg six times.

185

We looked at each other hoping she was teasing. We started laughing and wondered if this was some joke to see how gullible we humans are!!!

On the butt?
    Yes. All spankings open the root chakra.
Now?
    Yes. Use your hand, not a paddle.
Is there any other way?
    Wear cinnamon in butt?
No thanks! Standing?
    Lap is best but standing is OK.

We finally decided to trust her and reluctantly, I spanked Torg. My hand and his butt were stinging after. He said he felt a sensation all over his body that it was working. We had a good laugh even though it was awkward. It would have been different if we were in a romantic relationship!

She then gave us instructions on how to release him.

We were told not to publish all of the details, so it stays safe. There were eight symbols all together; one was an activator. Their meanings were: *mercury, life, listen, male/female together, eye watch, grounding, horizon*; followed by the activator. Apparently this was Pongust's real name in written form! We felt honored, since the elements don't give out their real names because evil Shamans can control them. Apparently there is more to unlocking his name, so his name is still safe from anyone controlling him, even us.

She gave us instructions for each symbol and at the end she told us to say aloud, "*Bring forth 'time out of time'.*" This would create the magic to release him after the drawings are washed off. We were allowed to take a picture before we washed them off, which will be kept secret. But apparently there were no real worries, as the binding must be pronounced in Gaelic to be effective, which we don't speak! Another fun fact:

Now for the first conversation in person with our long time friend Pongust, the Guard of Stonehenge!

Pongust, how are you?

Pongust: *Numb. Happens a lot to others who travel. Me, I've traveled only three times.*

We thought this was your first time to leave Stonehenge?

1ˢᵗ to leave Stonehenge. Happened when I traveled there.

Someone brought you in a rock?

Yes, when all land was one continent. The other time was when Gaia was molten.

We are honored. This must be very important.

Yes, you are. (Torg) I was around for all three dances. No metal touch me, please.

Why was aluminum foil wrapped around you?

To deflect watchers.

Are they listening?

Yes, be careful. They know the dance is coming now. You are safe now.

Are you going to dance?

Part, in the circle only. 6 men, 6 women.

Am I one of the dancers?

No. You need to be touching Torg's CSX (Center chakra). Get your team together.

Are you happy to be here?

I like my rocks.(Stonehenge) I finally can see you in person. Where to next?

We don't get to see you, though, just the rock your intelligence is in.

I don't get to see you either; just your earth suits. (A Kahuna Pahia term). Not the intelligence.

Do you want Torg to carry you all the time?

That would be best.

Are you surprised at how we look?

All you humans look alike. What are you doing?

Making cake. If you aren't with Torg, where do you want to be?

Touching his onyx.

What do you want to see?

Wherever you go. I'm easy.

Thank God!!

You carry onyx with you. We'll be very strong, we will glow in the dark.

Can I carry you?

If you get spanked! I like you.

Make sure Prescott and Mom are watching. Not all spankings are punishment.

Do you trust the djinn?

Yes.

Is he here?

Yes. If he rides my rock, we will rock the world. Burn a hole in your pocket.

Ryxi, you ready to ride me?

Not sure about getting spanked.

You humans forget easy.

Mom: Djinn needs a new name, Ryxi.

Mika? Does it fit?

Yes.

What does it mean?

"Window to the other side." Now a "she". Djinns change. Part of their magic.

Ryxi does much magic!

I don't know when I do it.

You need a spank. You talk to us, so you open chakras, Make cake. . .

You want a wand to zap things?

Could I?

Yes.

We don't see it.

Might. Need to open root chakra. Your crown chakra is already wide open.

What do you mean?

Very happy and sad for you. So much potential. I enjoy watching.

Where did I go last night?

With Me, MOM. You and dragons guarded for Mac.

When did he come?

3 A.M. no way he would wake a sleeping witch!

Did he see me sleeping through the window?

No, he's very proper. Waking a naked witch isn't healthy!

I wasn't naked. Did he wake one?

Did once, has scars to prove it. Came at 3 A.M. to be safe. On his way to a new assignment. He was freaked out by your messages.(notes on the door and mailbox) *Thought his cover was blown. My bad. I need a spanking. I go see Dad.*

Get a spank from him?

Hope. Ryxi - love you. I like Torg. Pongust not bad. Pompous troll. Reminds Me of Me! Open your root OK. Djinn has gone to the garden. Bye.

OK - Torg spank me! . . . I kissed the rock.

Put me in your pocket. Nice kiss. Some rocks like to be kissed . . Like the Blarney stone.

Is that real?

Yes. Heart of Ireland. A node!
You have more magic now, enjoy.

You mean after I get spanked?

Yes.

How often?

You? LOL. Have fun. I should travel more often. I want to go to Safe Haven, temples, LDS temple, Masons, all others. Make signs to protect Safe Haven. We protect Stonehenge by making signs that say not to touch the rocks.

He was talking about signs letting people know how to act at Safe Haven like being drug and alcohol free, taking care of the land, etc.

June, 2012

Mom, Pongust, Hilti, etc.

Pongust, did you have fun at Adam's temple?

Yes. Right two rocks.

What do you think?

Such a powerful place, now dust. Place honored from the Time of Patriarchs. I would prefer no noise. I had the same problem with Stonehenge. Michael's (Adam's) *Temple should be more sacred. Wish kids didn't scatter sacred stones and ride motor bikes over it.*

We took some of the stones.

You're OK. You honor them. You found some from some very powerful people.

Who?

Noah, Shem, Michael. They were in his tomb. Many women, not all stones are home stones. The white and black stone came from me, stonehenge. Other stones from Green Mesa. Many stones now are safe with you. Some you keep, don't give away.

How do you know they are Seth and Noah's rocks?

Their signature. Black and white - Mom and me.

Did you want me to go into the cave with you?

Yes.

Did I go far enough?

Yes. Michlin, the guard welcomed you. Onyx binds and draws closer.

Torg: Makes sense that the onyxes are from all over. There is an Enoch scripture about them traveling.

What happens when people stand in the portal at the end of the cave?

Nothing. Only Shamans can affect it.

Why?

Resonance.

Something they develop or just have?

Talent? DNA? Your niece, WOW!

You should take her to Green Mesa.

How does a Shaman affect it?

Splash. I read Adamic, Gaelic, English, Welsh, Saxon.

What do the petroglyphs mean?

Adam and Eve bound together for all time. Gods and creation. Adam and Eve traveled to this new home. Don't look here, they are gone. Several eternities. They were bonded before this.

190

Chapter 15
Keys of Onyx

Date: either April 29 or early June 2012.

Hi, the energy is off . New onyx from Adam's temple. identify them 1ˢᵗ.

Why are the onyx rocks important?

They are imprinted - Seth, Adam, etc. Noah is imprinted in another stone you put away. Feel the imprint. All ages have key people. You can use them as well with their imprints.

Confirms what I received on the property.

Keys of Enoch.

Flying?

Out of body.

A lot of people can.

Yes. Having keys makes it better. Keep stones on altars.

Should we carry them with us?

No Dumbo feather. Be careful.

What is the Seth imprint of resonance, or keys?

Patriarchal, matriarchal. Ability to talk to God and express the will to people.

Can't everyone?

Talents always better with keys. Surgeons are better with keys.

Keys?

Priesthood enhancements. The ability to act in the name of God.

Women already have the ability unless they are not being one with God?

Off track. Evil. Men earn the right. Everyone has talents. Not everyone can play the violin. Some have talent (scientific) ability to connect. You have resonance with Me. Most women do not. Having the right doesn't guarantee a connection. *Your people all have a right to vote; how many do?*

So some have more ability to connect and even more ability than others to connect on different levels.

Sure. Everyone has the right to play the violin, some have natural talent and can play better. Keys make both better.

What are the Michael/Adam keys?

Others are still in the tomb. This one is lesser.

Population - (replenish the earth). Fertility.

Noah/Gabriel keys?

Wash clean and start over. Cleanse, baptize, rebirth, etc.

OK, am I understanding this right? These onyx stones have the resonance of the original structure and/or person that carried or owned them. Their talent or calling energy still is part of the stone?

Close.

How could I be closer?

Be God!

Women's stones were mostly yellow or green. More about family than power. Not many at Adam's temple. More are at Eve's temple. You have one priestess warrior stone.

Should I find it?

Do you want to fight?

Not my calling?

No. Find and not use if you want. You have no use for warrior priestess keys. She was a Joan of Ark type. A captain of female spies. Iy was easy for a female to get into enemy camp. Sword in Stone drew power.

How did I become so connected to You?

You came out of darkness (the void) that way.

What are the keys of Enoch?

Command elements, translation (body changes), fly out of body.

Did he go to Adam's temple?

Lived there in the area. Went to Egypt to help with the pyramids. Lived on the West mountain.

Did it snow in the Winter back then?

Not much. Lived in patriarch valley with others. Mountain is not Seth temple. . . Seth mountain . . .

People say they've seen ruins in the Mountain Village?

Not a village. Millions of people.

Is the property we were going to buy near there still important for us to buy?

It's on a Ley line. Safe Haven, too. A sacred valley.

What makes it sacred?

Resonance to Us. The whole big valley is sacred. Safe Haven is one of the spots. Millions of people. Enoch moved south, took many with him, flood coming. Took the righteous people so they could leave. Fought giants. (Big humans, not element giants.) Broke away and left. Took righteous and built a new city. Evil did not translate, no resonance, died instead. No evil there in the city of Enoch. Evil people hated Zion, stayed away.

Why did Noah stay?

Cedars - wood for ark. His calling was to stay. Ark animals are real, not DNA. Many others survived, on mountain tops. Melchizedek and Shem are the same person.

Where did the water come from and where did it go?

Laws of physics, chemically bound, hidden in rocks, ocean, comet. Coral, yes. Chalk, etc. No worry, bad resonance.

Are resonance and vibration the same?

No, many vibrations together.

Have faith, my children. We know what we do.

June 3, 2012

I had given a speech at a retreat where some attendees were from a conservative religious background. I was very nervous about talking about women having the priesthood and giving blessings. My upper body even started circling and probably looked like I was possessed! In the middle of the speech she stopped me and I was tongue tied until I started talking about what SHE wanted me to talk about.

Brave Ryxi: women's priesthood wasn't easy for you to teach. Met fears head on.

How can events go better?

Expectations. Don't do events for a while. Matrix nights, yes.

You said there would be 70 people showing up. There were only 20 people.

All of you and My many others: guards, ancestors, angels, elements, etc.

Many others checked in. Even people who don't check in with us. Guards knocked Torg down. (Sparks - what they looked like.)

Why?

Tried to talk to him. Left when angels of the other side gave him a blessing. Torg should go back up the canyon and talk to the guard.

Alone?

Better not. Vicmor called the guards off.

Vicmor is a strong earth element over Mt. Olympus, down the canyon from the retreat. We've been talking to him for years!

Is Vicmor here now?

Vicmor: Yes.

What are they guarding?

Vicmor: *You need to study. A Sacred site. I was watching from a rock. I like dark chocolate better. Should have been naked. Rider brave, stress, beautiful, good, Goddess.*

It started raining and people started commanding the water to leave. Did we insult Big Water by commanding him to leave?

They just came to say hi. No worries. They're used to it. (being told to leave). A special "hi" to you. The Thunder was him saying "hi". Rained on your parade. You're all tired kids. Earned it.

Finn: *Torg , your feet won't heal until after you dance.*
Helps you leave and come back. B1 vitamins, laugh, walk,
sleep.
Mom: *Rider can hold Pongust, no spanks.*
I'm always with you. Let's talk Thursday. Hugs.

June 3, 2012
Do You talk to Leroy as much as he says You do?
No, more. He is your most sensitive student.
What is Your description of a Shaman again?
Grid tied power Walker. Healer. All Indigos and
Crystals are.
Quality, not quantity. The whole earth shook to one
Master. You rest. Must be well.

June 5, 2012
I stepped on an object left on the first step of a staircase in my
home and fell full speed down the stairs slamming into the furnace at
the bottom. I broke my toe and messed up the forefront of my foot
where the toes attach to the foot! It hurt so bad! My whole body
shook in pain.

**I felt my foot tingle while riding in the car. Who was
healing it?**
Finn.

June 10, 2012
Ryxi! Hugs!
Who is this?
Me, Mom.
I gave Torg a vision last night?

She was talking about Torg's vision during Matrix last night: Rider
and I walked the up ladder, hugged Mom, who spanked us and slid
us back down the ladder.

June 11, 2012

Come follow Me . . . Let me know what you need and I will do my best to deliver.

Don't You just anticipate needs and help whether I ask or not?

Yes, but not as much. When you ask specifically, it gives it more power, and I can help better/easier. It's magic! If you sing it even better!

When you go to Mesa Verde, make sure you tell Pauli she is beautiful and how grateful you are. She has done such a good job of guarding the space. She needs extra comfort and inspiration now. She is lonely and wants to give up. Make sure she gets a kiss and chocolate!

Pauli is the earth element guard over Mesa Verde. We met her the first time we went there. We keep in touch now and again.

My kingdom is your kingdom. Your kingdom is My kingdom. Many ways to have fun and still be on the path. My kingdom is joyful and painful. One and the same. Get used to it. It's worth it. Better to be free than be a doormat. Feeling is believing . . . confirmation you are alive. Makes life sing. That is why humans are Gods; they can feel.

Animals?

Kind of. Not as much.

Why are humans different?

DNA + intelligence + spirit = human.

Why take a humanoid shape?

Easier to get around. Energetics of spine being straight and vertical helps with brain function and tuning in.

Apes?

A lower form. Belong to other Gods with family. Helped with evolution process to learn. To practice Godhood.

Why let other Gods rent space here on Earth?

Because it was available. Wasn't time for our people yet. We were practicing somewhere else.

Why not here?

Started earlier with other planets to get the ball rolling faster. Not time to transplant yet. Wait until ripe. Meanwhile Earth was last created. Help younger Gods start to practice so be ready when time to start their universe. Helped Us, too.

How?

To become whole. Needed the energy of human. Plants and animals evolve and needed to be used to human energy or would be off balance when humans came later.

Sacrifice.

You know it makes things stronger.

When in the sweat lodge and it's hot, I say it's a sacrifice for all living.

Yes, Makes the ceremony stronger.

So I complain about all the work for events, when the sacrifice of the work for events make the events stronger?

All do.

So I'm being a baby.

Mature more of course. Even Jesus complained.

Jesus?

Cup too bitter.

Was His cup too bitter?

Not if you want to become a God. All Gods give the ultimate sacrifice. Father and I finished after death. Jesus before his death. Torg's dad did his sacrifice before. He didn't get whipped, etc. Everyone's sacrifice is different. All don't have to die. Give time, talents, resources. No two people alike.

I guess we are working on that.

Oh, yes.

So I shouldn't complain so much, and think of what I'm doing is becoming a God.

Don't get a big head, get ahead of the curve. Take time, take stock, take rest.

I'm going to wear a short skirt later. Mini toga.

What color?

Hot pink.

Dad's favorite color?

Don't change color.

There you go!

She danced to "Yellow polka dot bikini".

Should I wear shoes?

Hot pink high heels.

OK.

June 12. 2012

When you go to Mesa Verde, think about how much you like being in My service. It will help others feel the same way. Send out to the Grid . . . Service to each other, the kingdom of God. When I come to you, let Me know your needs.

I need my foot healed, car keys found, details for Mesa Verde . . .

Let Me into your heart so I can be Myself through you. You will have my countenance.

Don't I already let You through?

Yes, but could be even better! Still holding back. You are My shining star! You will help bring to pass the immortality of man.

That sounds like your job.

Yes, through you We can do it together.

Yes, but won't everyone become immortal eventually?

Yes, but could be even better. All intelligences are immortal. Spirits are not. People become immortal with all three and become Gods! Body (physical), *Mind* (intelligence), *Spirit* (spirit matter) *together as one become Gods.*

Of course when they develop to their full potential. So when You talk about eternal, you mean people with all three developed into a God?

Yes. Much more effective in the universe. Our release from loneliness is indescribable. Everything is automatic, but We can influence it with extra nudges and encouragement.

I thought about how I like to help other people not in my family and how She talks about things "not on Her table." I realized She means responsibility and stewardship. That helping others in other families is nice and all, as long as your own family is taken care of

198

first. She says there are other Gods over Hopi and Aussie Aborigines and They help as long as They have extra time and energy. As I was thinking about this concept She read my mind and said, "*Well said!*"

I tried to take on more than I can chew sometimes.

Gods do that a lot! Can over serve and get burned out. Can under serve, too. Much balance in my kingdom. Many are called but few chosen / show up / choose the path.
Anything else?

Love you and I still spank! This will hurt Me more than you!

June 13, 2012

You rock! Ryxi in tune with Me! (Danced infinity)
How? Specifics?

No box, a package deal. Safe haven, Green Mesa, blonde, cute. Don't over think.
You're very excited, so I thought something was not normal.

You not normal.

What's been going on with Torg? Are bad things trying to take us out. (Every where Torg went, cars were trying to run into him like he was invisible.)

You are in danger because you are doing good.
I'm surprised we don't have more guards.

You do! You didn't get hit! Danger all year because of calling.
Have they always known my calling?

Yes, ramping up. Tried to get you to quit. Made you fall down stairs.

I tried to get permission to bring my niece to Mesa Verde, my sister doesn't understand what we are doing.

Always hard to get people to wake up. Many people were told of the Flood. Only Noah built the Ark. Mesa Verde not just an added idea. You are in tune. Very important. Torg, talk to guards. Another Shaman, Ellen, will too. Another group will be working the lines. Other Shamans besides you are connected to Green Mesa. You are the strongest. Your team opened the portal to the Grid. If Jesus didn't do his calling, We would have asked someone else. Most other Shamans don't even know you. Their loss. You healed Slic and others felt him after. More come now that Mesa is open. Because of what you've done at Malibu, Tikal, Safe Haven, etc. more Shamans are coming.

WOW.

Of course. All you do has an effect. If you are happy, places get better. If you are angry, places get worse.

Ryxi's job is Ley lines and portals. Torg's job is knowledge and power.

I thought his job was the guards.

Guards are power.

Safe Haven?

Take the djinn and put it in the dragon shed. It will make the plants grow. He is a good boy, now. Bye. Talk before you leave.

June 2012
Summer Solstice

Solstice was amazing. We ended up doing ceremony in an actual ancient Mesa Verde Kiva! We brought powerful local Shamans with us and an amazing musician and Didgeridoo player, Leraine Horstmanhoff. She played and sang while everyone chanted as we set up the ceremony. We performed the Male/Female ceremony and then our Peruvian Shaman friend performed a despacho ceremony. As soon as she started the wind started blowing. It was obvious it was blowing for us! It was amazing! Unfortunately, Torg went to bury the despacho and got caught digging by a ranger and was given a ticket. When we got home, Tizan, the earth element from Chaco canyon came in first.

Tizan: *The wind that blew at the kiva is here. It followed you home. I liked the incense you burned. The red one was best. You guys rock!*

Wind element from Mesa Verde: *I like mountains, fire, heat (thermals).*

Mom, what were the highlights?

Ticket no problem. Wheels in wheels. The big picture, wow.

Was it worth getting the ticket for the big picture?

Very much.

Sacrifices that had to be made?

Close, no box.

Torg's life will soon be more busy.

Was doing the ceremony in the Kiva more powerful?

Yes. Sacred wheel in wheel. In the sink would have been better. Earth cycle stronger because you were there.

Gaia: You tickled Me. Cycles are important, cleans chi! There are many sizes of sinks and fountains. This one is medium. Big ones run from nodes to nexuses.

Do all have sinks and Fountains?

28 have a node to a nexus. You know feng shui?

Some. Were other nodes worked on for the solstice?

Oh yes! (laughed). I go. Others need me.

Was the boulder near our campsite a portal?

Yes. Old ceremonies. The scratches were graffiti.

The pine cone mountain was shaped, not built (Mesa Verde). They took dirt off until it was shaped like a pine cone. They didn't pile dirt to build it up. A Pyramid shaped like a cone sends magic up. Energy exuberance, wow to universe. Creates an antenna. Listen to whoever does ceremony on top.

Ceremonies were performed all over Mesa Verde. The sink is unique.

They're talking about a hill near our campground that almost looked like a pyramid or a pineal gland.

June 24, 2012

You rock, happy is me!

What is biggest rock?

You mixed ceremonies!

We had Native American, kirtan chants from India, Peruvian and Australian Didgeridoo for our ceremonies.

Yes, big challenge! Now I understand why You said "only friends."

Conflict, not much. Cute ranger. Ticket still no problem. Torg, your Mom is happy. 1.) You danced with the guards and stayed safe. 2.) Helped Ryxi. You were fine, didn't hurt much. (Torg took over ceremony.)

The Djinn is happy to be at Green Mesa, lots to grow there.

When I was dancing in the Kiva during the ceremony Leroy saw you dancing with me!

Yes!

Who else saw you?

Shaman ladies. They now respect you.

Torg will go soon to Island (Philippines), talk to Shamans there. Take a small kit.

Leraine is awesome!

Gifted, yes. I told her to work with you. The shaman ladies are on their own path. They will join you for a while. This was the most powerful ceremony yet. Always improving. Leraine listens. All your team did wow.

Did you see Masau?

Was that the odd feeling I had?

Yes.

At one point while camping I felt an odd sensation. I told Torg I thought it was Masau.

Wow, I got it!!

Masau heard. He knows the way there. Masau's ancient people lived there. He laughed at Torg dancing with the others. Torg needs to go see him. Go to Hopi sometime.

June 26, 2012

I don't remember why I asked this question. I think I was worried about people who turn to prostitution for drugs.. Or maybe I was wishing I could have a partner. Not sure.

This sex thing can really mess with people?

Yes, a powerful tool for good and bad. Many fall because of it. Many ways to be twisted and confuse people. Much good, too. You have to use it very carefully and have fun, too.

What now?

Come follow Me.

Yeah, You say that a lot.

It's true. May I have this dance?

Of course.

Make sure you stay healthy and listen to Me. Most prophets forget to pay attention to their own needs. Best to be strong. You need your health to work for Me, so pamper yourself now. Get well! Greens, dance, make out. Lots of water, lavender, salt, etc. Really go overboard! Pamper your feet. LOVE YOURSELF! Get organized. Sleep well. Good job in resting for the weekend. High 5!

June 27, 2012

How do I help Paul, he's going through a tough time right now.

Hard call: tough love or coddle. Hard to provide shelter without making someone weak. He needs both.

June 28, 2012

Men and women need each other. Like horses need rein?

That's random.

Life is random. My mission is to tease and to teach. Funner that way. Anchors lessons better. You should try it! Many ways to teach, fun is best. Although pain works, too. Consequences of pain make people grow. Inspire them with fun, spank with pain. "Love you."

You must know that I am tired of spanking. Would rather hug! Funner! Life is full of spankings. Many more to come, get ready! On your mark, get set . . . spank!

Many more hugs, too. No worries, life is both. Me too, still. Different spanks now. More like experimental spanks. Many called, but few spank.

So I should just get used to it?

I did! (Visual of Mom and Dad spanking each other in fun.)

Let's work together and be all right! (Bob Marley song in my head.)

Get well, kick in butt whatever you have. Eat garlic.

July1, 2012

Djinn is fine. Hot day there.

Cleansing by fire?

Yes! Gaia is heating up.

Where does it start?

At the very beginning; "doe, a deer, a female deer". . .

What was on Moses' tablets? Before he had to go back up the mountain and got the 10 commandments?

The Adamic law that Christ brought back. Love thy neighbor.

Why not before?

If given, expected and judged by it. Long time in Egypt being told what to do, couldn't think for selves.

Now lots of people are good and don't need to be told what to do?!

Yes! Adam's law. Personal choice, sovereign. Responsibility. Moses' law is clearer and more defined. No responsibility, obey yes, no thought needed.

Do good to your enemy.

Some people think open marriage doesn't hurt anyone . . devil's advocate.

The Devil doesn't need an advocate. Has too many! Wrong thinking, leads to wrong actions under Adam's law and punishable. Under the Mosaic Law it is hard to think wrong, everything is black or white.

Bob Marley did bad things and still did a lot of good.

204

Bad people do good things. We judge the whole person. I did good and bad, too. Thank My God the package was good enough! If you do bad once, it won't destroy you.
What about twice?
No, seven. One good deed doesn't save.
You people are enslaved the same as Moses' people. Smart people see the box. Taxes are rules by the elite. You are enslaved now like the Israelites. Schools, speed limits. Need new laws.
But stop when? Start where?
Spirit. The man cares only about the man. You work for your house 50 hours a week for what? Work for life instead.
We're slaves more than Israel was?
Your slavery is mind stress. No box. Their kids were taught not to a rebel.
So are ours.
You must trust your God to know what We are doing. Would you let a baby show you how to drive a car? The human race doesn't know enough, wise enough, or mature enough to do God things. Sometimes you must have faith. Your kids had faith to go to school because you said so. Now they know you were right. Please try to trust.
It struck a cord when You said government oppresses. I think some churches oppress too.
Egypt ran their church.
Leroy: Do mermaids exist?
Definition?
½ human, ½ fish.
No. There's a high intelligent society living under water. Not fish, whales, or squid.
What do they look like?
2 legs, 2 arms, gills. Some have a ridge down their back. No hair. Big eyes.
What do they eat?
Whatever they want. Blue blood. Better for oxygen down deep.
How do they live?

205

Long talk. Don't have to protect from weather.

Do they hurt humans?

Only if attacked, fight back.

Are water elements mad at us?

Big Water is sad. Too wise to be mad. Leo bad a dream.

Leo: Do I have a Lion animal spirit?

Of course. Your name is not an accident.

Any humans ever see water people?

No. Been here a long time, lots of them. Came from another place and stopped.

How?

Portal. Big story.

Neanderthals?

Nice people. I miss them. Not ours.

What happened?

Did not go through a portal or die out. They bred with others. Are still here.

Bigfoot?

No. Gentler. High order of primate, smart gorilla.

Any on Mars or Venus?

Venus not much life.

Mars?

Worms and stuff.

Will humans meet the water people?

Yes, in the oceans. Several types of water people. Not all alike.

What happens when Hopi or Australian people mix with your kids since they are from other Gods? Or anyone for that matter? What God do they go home to?

They can choose.

Cheyenne or Cherokee?

Mostly. Breeding muddies gene pools.

Are you OK with interbreeding?

Sure. Mutts are tough. Can't breed with water people.

Why bring Adam and Eve here?

Needed here.

Will Gaia heal herself?

No time. You guys will. Would take a million years.

Is that why we teach?

Yes. Heal her in a thousand years.

Did You choose the Shamans directed to our class?

Most.

Did You choose Leroy?

Of course. Leroy used to listen to Me more. Didn't know it was Me. Now you usually listen. Little kids don't ask, they just talk.

Any suggestions for the 'out of control' kids in the class I teach at church?

The Flood worked for me. Ha Ha

More fires are coming.

Water people laugh at humans when they drag cameras and think you know something.

Can we work with them?

Not yet.

No going out of body. You need to rest. No cumin. Ground with oregano.

July 9, 2012

Who's here?

Pongust, Pax, Milni, Hilti.

How are you all?

Rocky. Ha Ha. Been watching you, sister.

Are elements getting in my old room?

No. Ancestors, guides, etc. Residual energy. New tenant sees afterglow.

Do they visit the new room now?

Of course. Think of glow in the dark. The sun is gone but still shines.

MOM: *Hi Kids!*

Jesus said the meek shall inherit the earth. What is your definition of "meek?"

Teachable and adaptable. I don't want pompoms.

Pongust is pompous, is that OK?

He's OK. Gaia is changing. Stewards of the earth must change, too.

Or go away?

At least fill paradigms with change. Meek does not mean submissive, slave, or servant. Except a servant to me,ha! Those who serve me are meek. King David was meek at first. Gandhi was one of the greatest leaders; meek, yes. Joan of Ark, meek, yes. Burned. Came home.

What about people who burn? Do they feel it?

Most don't. Varies. We usually take before pain. Joan of Ark was happy. She knew she was coming home. Don't cry for them. They are safe. Feel sorry for their killers! Living is more painful. Easy to cross over. Ask Torg about living. A slow death hard. Torg's dad's death was slow and painful.

What about euthanasia?

Sometimes. Torg's dad's mission was to die slowly. To teach his Mom to serve him. He served her all those years. She learned to serve him. He's OK with it.

Why did my grandma stay so long after grandpa died? She hated to be without him and was so miserable, wanted to die and be with him. Did she have a mission?

No, his mission. He needed to be alone without her for a while. Won't say why. Just that he needed time alone. Learn later.

Torg had a dream about a black cat that died in the Sweat Lodge so Twiggy could have more chi to live longer.

Twiggy is a very special cat.

Do you know where Bubba is? (My cat that disappeared.)

Won't say. Doing good.

Calling somewhere else?

Won't say.

Do you think that I hear things not meant for me to hear because our minds are one?

Yes, resonance.

Like when I got the location of the Ark of the Covenant?

Yes, dammit!

July 15, 2012

Hi, still Godly.

Journey last night to Keebler. Long away place, not ready. Other dimensions are waiting for you.

I went to do an exchange?

None of them are here yet. You made a good impression. They will come soon because of your good impression. Channel opened, you fill the need and went. You were talking to them with a guide.

How long was I there?

About 1 hour, talking to guides and left. Said "Hi".

Why call them Keebler?

Need to keep sweet during Millennium.

You need rest. Torg is sick. Torg journeyed to an empty space. His body needed to be alive, came back. Losing cohesion, falling apart. Shaman process, make feel sick, slide.

You will have an appendix where many people can write. (In the book.)

July, 2012

I've been so tired, is going out of body and visiting other places important enough to be tired all the time? *Everything is your choice. I asked you to rest. I suggested only a few sweat lodges. You chose to answer a call to another universe. We are grateful, but please don't be mad at Us!*

I don't know what to do to feel better!

Rest. Do what you have to and take care of your body. Relax!! All in good timing. Yes it is tough. Submit to the resting. You are a God and are doing many things. STOP being hard on yourself!! Try intentional resting!

July 16, 2012

Hold Pongust to keep you grounded so you don't fly. You need rest.

Is my flying beneficial to the universe?

Yes, it helps. But fine for a while.

Many Prophets make mistakes. It's hard to know for sure they work for You.

209

Just like you and Torg and many others. Even with prophets, we choose the best person for the job, flaws and all. Mohammed, Gandhi, Da Vinci, Confucius, Newton; all had major faults, but did a good job for Us. Still the best person for the job.

As long as they do a good job, they can do anything they want and be forgiven!?

They are not forgiven! All evil must be paid. All good must be rewarded.

So I hear Luciferians think that as long as they do more good than evil, it's all OK.

WRONG! Many of the Prophets still have much to pay!

I thought you had to heal to resurrect. Can you resurrect and still owe (need to pay)?

In some cases, yes. Very rare, less than a dozen.

It helps to know that You are acknowledging that prophets are human and can be wrong.

Chapter 16
Truth hurts

I'm sorry, but I don't like some of the ways things are being run here. I hope that when I am a God I can do things differently.

Would you allow your kids to tell you how to run your business?

No, but I'd let them tell me how they feel about it!

So do I. I'm not upset! I thought the same until my resurrection! I thought I knew more than my Parents. It's hard to be patient.

I had some tough trials at the time and was falling into a deep depression. Mom had answered some personal questions that I didn't like the answers to. I wasn't sure if I ever wanted to talk to her again.

I had also received some sad news and my foot was still hurting after falling down the stairs, among other reasons to be depressed.

Finally I decided to limp, broken foot and all, to the park and lay on the grass hoping to feel better. I asked the earth to take my depression and soon felt the depression leave my body and go into the earth. A huge load was lifted.

July 22, 2012

It had been a week since the earth healed me at the park. I was still sulking over my last talk with Heavenly Mother and reluctantly went to the park anyway. I still had a job to do for the kingdom even if I wasn't really happy right then about the kingdom I happened to end up in. I had a speaking engagement in 3 days. I sat under the trees for a while, not excited to talk to Her. I finally bowed my head and reluctantly prayed.

Thanks, to whomever healed me this past week.

Many helped. We need you well. Let's move on to the next step of your journey.

Who helped?

Gaia, of course, trees, angels, Finn, Me, etc.

I'm trying to keep my enthusiasm strong. Why am I so tired?

Makes sense. You are doing a lot of work that takes energy. Work you don't understand yet. Energy you don't understand. It depletes the physical body. Being an earth connected grid walker is a hard JOB! Then there's all the other stuff!! Dimensions, happy times are your way soon! Get ready, set, go!

She gave me a vision of a party with noise makers, balloons, etc. and heard the 80's song, "Celebrate good times, come on!"

I think I'm not going to ask any more controversial questions until after my job is finished.

I understand and went through the same thing.

I like my la la land world. I don't want it shattered by reality. I'll keep my rose colored glasses on until my job's over or otherwise I might quit or be sick and depressed.

Suit yourself.

I read a quote on Facebook I really like, "The truth will set you free but first it will piss you off."

I will add, "depress you," there are harder things than the truth, but I don't know what. Ask when you are ready.

OK. So when I know I have a week to be depressed, I can ask all the questions I might be depressed about the answers to?

Ask one at a time.

I should take a month off and go in the mountains where I can be healed as I go through the trauma of truth.

I sat in a hot pool.

On your earth?

Yes. Many more later, big picture comes in small pieces. Good you let it go for now. Be more flexible and it won't hurt so bad.

I thought I was.

Not enough. Could be more. Still working, through. Still doing your job. Would be more fun if you were strong. Not so sick.

Good you let it go for now.

Oh, well. Moving on . . .

Yes, I need you happy.

How did my depression go away?

You, Me, the trees. Ants, too. All of life. You are the Ryxi! Poor Goddess, reality hit you in the face...Butt. Reality has sharp shoes!

Any suggestions for my speeches on You?

Remember central themes. Always come back.

Themes?

Stewards of the Earth, Ley lines, prepare for Millennium, male/female in balance.

Heavenly family - We are one - You've always prayed to Me even though saying, "Heavenly Father" while praying because we both answer.

Time for Divine feminism.

Always smudge, pray and protect before and make a circle for protection.

Book?

Not ready yet.

Is the book not ready, or is the world not ready for the book?

Book and world. Do your best first. When the book is ready, publicize as much as possible.

My class of kids at church are a challenge!

You see what I go through? LOVE YOU!

Yes!

Relax tomorrow

Late July 2012

Torg and I led a workshop on Shamanism and another on Heavenly Mother. It went well as far as we were concerned. People seemed to like it.

Hi, (She sang)

Wake up and do something more, than . . .

Relax. I am with you.

Vision of her dancing

213

Can I have this dance? Watch me make miracles.
Wow, oh wow.

I guess You're happy. Highlights of the day?
You! You are no blonde!

I got in most of what You wanted?
ALL!

I wish I could hug You. (I felt a tinge of good energy) OK I feel it. Ah...
Mess up my makeup.

I felt honored I made Heavenly Mother happy enough to make Her cry.

You two in tune. Hard melody, but scary. Hard to face fears and go forward. On page, on target, on time on my list. You are one of the small elite that is following her calling. Many are called, few are chosen, fewer show up. Still fewer walk their talk. You are a true daughter. Don't let it go to your head. LOVE YOU! You a true Goddess today.

What's different about this one than other workshops?
You were more visible. Thought you would back out. You are now published, more credentials, add to resume. Sure did Ryxi. Turned some heads. Ripples to high surf. Catch that wave!

You wait. Prophecy fulfilling. You are part of prophecy. Restoration of John the Revelator. Fall of the Horsemen.

Which are we working with right now?
All!

I went to some of the other workshops and spoke really out of the box?
Fine. You shook them up. Some will listen.

I'm not used to being around people who don't believe us.
Some did and more will.
My toes are blue. I have long lashes.

Torg's feet didn't hurt during the workshop.
You're welcome. Only for conference.

This is just the beginning. A long road ahead. You're fine. You have good shoes.

What do you see down the road?

You being great. Much fun.

If it's fun, it's not that long of a road.

Long, fun road!

How many people will be in the audience down the road?

Millions.

At one time?

Random. Internet viral fun. Best I ever created.

You created the Internet?

Yes. Get the word out!

Chapter 17
Power Blessings

July 25, 2012

In the beginning was the Word, and the Word was the Internet.

I have a great team. I hope they keep coming.

You get a plaque. Blessings wow!

One girl at the workshop cried when we gave her a blessing?

Yes. The blessing was true, so she cried.

I was scared to give blessings in such a public forum and with people from conservative religious backgrounds. It was also hard to give answers to questions and people being upset with the answers. Kinda like I get when I don't like the answers!

Big steps. Bad shoes. No worries. You spoke truth. Truth is painful.

It would be nice to hand truth on a silver platter.

If you figure it out, let Me know.

My role is to piss people off with the truth! Either love me or hate me.

Like all prophets and traffic cops! Tomorrow more laughs.

July 29, 2012

We're early

Early is fine

We could hear my son playing video games in the other room. Between him yelling and the game sounds it was loud.

We didn't have those kinds of games on my earth. I was born after electronics, but we got rid of them. They got too smart and took over.

Video games took over?

Yes. Happening here now.

You're saying electronics are taking over here now?

Intelligence. No worries, you will back away like We did.

Wouldn't they be an intelligence with an electronic earth suit like ours?

No, you misunderstand. Programs become smart, self aware, while programs.

There were angry feminist women at the work shop. How do you feel about that?

Blocking their own progression. Angry, not scholars, not listening to Us. No Holy Ghost in it. You help?

Did you like the comments I made?

You were right! No second guess.

I cry for My children, they do not listen to Me and they hurt. Men and women are at fault. If My authority and Dad's isn't used right, it's better just to join a social club. Men need to support Me. Women need to support Father. Ignorance is stopping both. This now must end! This NOW must end! Will you help?

I guess, if I know how.

Yes, you do. Many ripples. Many hearts broken because you spoke with the spirit. Many heard you testify, you pray to Me and get answers.

Dad's authority is much like a policemen. Go lay down the law, make sure everyone follows.

Policemen are usually good, but sometimes they like their badge and not let people be their own selves. Male priesthood is almost always good, but all they know is Dad. They do not understand My authority; they were never taught. Neither were taught. Not enough. You fix now.

So I'm supposed to teach about Your authority?

You must know this. You were a waitress once. If you came to the table with food already on it, would you offer food? All must ask for righteous things and right things. Men's authority governs. Women's authority governs life.

What do you mean by Men's authority governs?

Men's priesthood runs things.

That's very general.

Women run men.

218

Teasing?

No.

So in a family situation a man's priesthood would be...

Running a car. Men do not govern over the home, it's joint with female.

So when it comes to being a God, women are over life and men run things like making sure things get done?

Men make houses, women have babies. Men build churches and women sing. Men make planets, women put life on them.

July 30, 2012

In the trees:

What happened to My thoughts about you being My mouthpiece?

What do you mean?

You didn't write down what I said about you being My mouthpiece.

I thought I did!

DIDN'T!

Maybe I don't want to be a self-proclaimed prophet.

You're not. You're proclaimed by Me. People feel it and "know." No worries, I will testify to them that you are my mouthpiece. YOU ARE MY MOUTHPIECE. Take up the sword of truth! Now is My time. You have been called and prepared since the beginning. No shrinking violets, please!

Was I shrinking?

Not really. But you will be called to do even more.

I felt uncomfortable with Her calling me a prophet, I just wanted to write a book about Her. I didn't think I was signing up for this! I was pretty naive I guess. I looked up the definition of a prophet and got::

He or she receives a direct message from the divine, which is meant to be communicated to others.

He or she is able to somehow tap into divine knowledge and make predictions about the future of the world or about individuals.

I did not feel like a prophet. I had not lived a perfect life like a

219

prophet. Maybe I got the job just because I volunteered? I think everyone can be a prophet.

I think, even if people don't pray to You and if You are in their heart and they are listening, the "Meek" will hear You.

Resonance. Not from Me today. Heart from earlier. Comes when understood. Will happen more when you do your calling.

You have really opened their eyes. Good job Ryxi. I love you and am so grateful you are stepping up. Please know I understand and am with you and all My daughters and sons!!

The more you talk to Me in the trees, the better connected we are and the better answers you will get.

OK, what do people do to pay for their sins?

They set up a plan of action. They serve to erase the repercussions of their sins. Like being guardian angels for their posterity to fix sins of the father. Once they have made things right, they are free to move forward. Kind of like the Karma Show you watch.

What about Jesus' atonement for sin?

He just opened the door to make it possible. A sinner feels shame, are lonely, out of place, have no peers and relive the sin. We remember not if they earn and learn. Recompense to pay for sins. Otherwise door was locked. Never compensate. Still need to make amends. Jesus overcame the space of what We can do and what We can fight. Imperfect beings can't create perfection without help.

What is perfection?

Me, except on Thursday. No good way to describe it. I'm still getting better. Perfect and finished are not the same thing. I'm a perfect student, not a perfect encyclopedia. You kids always surprise me and come up with new stuff.

Ambiguous answer.

Perfectly.

Sometimes we think we have sinned when we have not?

Yes, must pay for that, too.

July 30, 2012

Will I present again?

Yes, you were well received.

Good feedback?

Yes. Blessings. Yes. Spirit very strong. I told others to come to you for blessings. WORKED!

My tree talks better than most in-tune spiritual people.

No concern, learning curve response.

We've been talking a lot on Mondays. Should I come every Monday too? Would you be available?

Information in the universe is always available to pick up on. Like voice mail. You have lots of e-mail from a long time ago. Trees remember, but are unappreciated.

Torg wants me to remember to talk to You in the trees before he and I have our sessions.

No worries. You are Gaia huggers, not tree huggers. The new tree is happy. Sings a lot.

Resonance. You need to take a salt bath in morning to balance your resonance. Torg might leave for a long time in a week or two.

August 6, 2012 (In the trees.)

My ways are always more simple than you think.

Huh? Seems complicated to me.

Not really. Just do what needs to be done, even if it is nothing. Eventually everything that is important will be taken care of. No need to stress.

How do you know what needs to be done?

Started singing: Stop, look, and listen. When you are in the flow, you always know.

Later with Torg...

Who's here?

All of us: The P's, Kyle, Keats, etc.

How are you Picci? (Machu Piccu)

Feeling peachy.

Why did Kucho cancel Mesa Verde? (A visiting Shaman from Machu Picchu who was supposed to meet us there for ceremony.)

Over scheduled, tired.

How are you, Kyle? (Leprechaun)

Wondrous. The council convened. The show last night (Nova - multi-verse.) is very close. I come from one of the muti-universes. The guards, too.

We watched a really good show on the Nova channel that showed the possibility of many universes besides our universe and how there could be many besides ours. It was exciting to watch and know they are right! It helped us understand our friends from other universes and helped confirm they are real.

Are you from another universe or dimension?

Define "dimension".

Layers within this universe?

Close. We're from another Universe. There are portals. Very big talk. Your scientists do not know enough yet. Learning very fast now. Most since Torg's calling. Hundreds of universes have portals with you now. We're in this universe now, but came through a portal from another universe to this one. You can too. Many missing kids go through portals to other universes.

Aren't they scared?

Some are fine, some not.

Can they get back?

Some. Some come back in a different time.

Do they grow up and come back?

Not always. Sometimes only their spirit or intelligence comes back. If they get lost they can be recycled back.

Why can't we see them?

You can. You have. They look like you. This is the most common type. Portals sort by resonance.

Why can't I see you?

We're inter-dimensional while here. Some can see us. It's like when you spin a bike wheel fast, can't see the spokes.

I'm worried about going and not coming back.

Take a guide.

Did I?

Yes, a friend.

Why am I tired and depressed?

Sun spots - energy flowing through you from the Sun.

Torg gets a lot done even with the sun spots.

Torg has resonance with the Sun.

Of course, he's a dragon. Pongust, how are you?

Grand and glowing. Mostly want more of everything.

What is your favorite?

TV. Do more stuff, anything to make more full.

How are my animal friends?

The Squid is home.

Are the Whales mad I haven't been able to meet them in person yet?

Disappointed. Not mad.

Were they waiting for me to show up?

No, doing their jobs. Did you wait for Kucho?
The Millennium started 10 years ago. Will peak in ten more years.

Approximately when Jesus comes back?

No.

Do you know when?

No, He won't say.

Torg's headache?

Mint, vent space.

August 7, 2012

Mom:You teach My mysteries.

How are You?

Me? Wow!

Did we do something good or are You just in a good mood?

Both.

223

Why are you in a good mood? (I wondered if it had anything to do with Dad)

Life! Long game, soon to climax. LOVE YOU! You teach my story, WOW!

Teach about your time, the new Millennium?

Mostly.

Have You been watching me write the book?

Yes, inspiring you. Change words, yes. Concepts, no.

You mentioned before that men and women should be using their priesthood together and giving blessings together. You said that is how Satan will be bound during Your Millennium. Again, what is your definition of priesthood?

Authority to act in God's name...God channel. The God channel is opened for men as needed and earned. Women have their God channel open always unless they shut it off. Being unrighteous can shut it off. Taking power for self. Same for men, channel is shut off when unrighteous.

You've mentioned before that women already have their priesthood because their priesthood is life and they have their authority over life just because they give birth. They have no need to be ordained. Men receive their authority when they are ready to serve and administrate for God.

Truth is truth. Not too much or it is rejected. Like a genie in a bottle. You need diplomacy. Do you know Will Rogers' quote? Nice doggy.

I'm concerned about getting money for blessings. I want to make sure I'm getting the right info.

Yes, make living OK, give to everyone freely and trust Me. Don't turn anyone away if they cannot contribute.

Lately I've been selfish. Not going around saving everyone. I hope I'm not out of balance. Have I been selfish or just taking a break?

Yes to both.

I think I made a big shift when I realized I can't save everyone and people need to save themselves.

I can't.

You must need to be detached to not get too depressed.

Not detached, pragmatic. Some people need to go to hell!

Where is hell?

Over there. Not a place, a condition. Lots of dentists there to treat gnashed teeth! No mascara, too much weeping! Lot's of ear plugs for the wailing! Funny for us who are not gnashers.

This sounds great for comic routine material!

Heavenly Mother stand up! How many Gods does it take to change a light bulb? God's light never goes out. You could go on You tube with Mom stand up routine!

People might not take me seriously!

No angry God on throne! Blue robes, hair auburn, white flip flops, rainbow sash.

Sounds conservative for you. Are you getting ready for a meeting?

Yes, take Dad off guard.

Ambush?

Ambush too hard. (She started dancing...)

Are you dancing because Dad is coming?

No, because you are a good kid. Wear happier clothes, bright. (I had been wearing black to look thinner)

Romance is the essence of the universe, family and water polo. Time's up. Let's talk tomorrow.

Hope You are in a good mood tomorrow too.

Depends on Dad.

I knew You were in a good mood because of Him!

Bye!

Aug. 8 2012

Long time no see! (Dancing)

I think She's happy.

Dad won me.

Were you in a disagreement and He won you back?

Not your business.

You brought it up. Anything more You want to say?

I'm feeling good. Playing with my skrog.

Skrog?

A hairy dragonfly. A very pretty resurrected thing from here. Violet membrane and wings, blue hairy body with two sets of wings.

Is it's hair blue too?

Yes.

Sounds pretty! Like the movie Avatar. Did you like that movie?

Yes, it was inspired.

Is there a world like that?

No, ideas from all over.

A combination of different worlds?

Yes, can't wait for Avatar 2 to come out.

Do you like sushi?

Live fish is best. When I was human I liked raw vemain. Looked like a centipede.

Yuk!

You eat crab. . . My nails are rainbow.

You're a nut!

Cashew?

Why Cashew?

Bent and sweet. Everybody loves a nut!

Wow, Dad really rocked Her world!

Other way around!

It's one of those nights!

One of those lights!

What language is Ryxi?

Not a current Earth language.

What is the name of the language?

Not around now.

Who?

People you don't know. Your name isn't random.

Why name me in that language?

Your name is power...it's about resonance.

August 12, 2012

Hi you!

How are You?

Happy are Me.

My class with kids went better. Were You there?

Always. Less brats. All kids are brats!

Even me and Torg?

Duh. Love you anyway.

I've been feeling so cold and disconnected to people. I'm usually outgoing. I just want to be alone. It's weird.

A phase. Soon will pass. Then you hate people.

I don't feel like I need to save everyone!

Just part of the universe that's worth it. Just this one. Maybe one more. Not Chicago. Or Amsterdam, the Bronx . . .

Do You watch the Olympics?

Synchronized swimming. Pretty, female.

Leroy question; Why did the Mayans leave their cities?

The culture changed. There is always change. No more bad priests. People were tired of big rock places. (Pyramid temples)

Like our hippie friends who move to the country and grow food?

Yes. Priest craft was not working. Your society is not working either.

What happened to the priests?

Went to work or starved. Getting paid is OK, just not overboard. They went overboard, were priests for money and not love. Most church leaders are good.

Were they mostly male priests?

Women too, bad with power. One of the tests. Hard to pass.

Did you have a meeting last night?

No. Pretended to be a meteor for you kids, fun!

I remembered I saw on the news and everyone was talking about the meteor showers. And to think one of them was Heavenly Mother! Too funny!

Leroy had a dream of evil aliens.

There are some. All planets have evil.

Do You have to protect our planet from evil?

Good and evil are everywhere.

Humans from other planets come here to do evil?

Their agency. We only allow enough to test you.
So there are evil aliens!
Of course. Keep to a level we can manage. Fools.

I remembered she said there weren't any evil aliens so I was confused, but when I read it again I saw she meant from other universes, but now she's saying there are some from this universe.

Do pockets of universes each have a God?
Yes.
Torg is in sync, understands this universe stuff better.
He's a scientist. You, Ryxi, understand artistic stuff more.
When Jesus goes surfing, can sharks bite him?
Silly question...can't hurt a resurrected being. Shark won't try and couldn't if tried.
What is Father's hair like?
Not artistic. White. Shoulder length. He wishes I would grow up. I wish He would loosen up. A good balance. He spanks Me, I caress Him.
Does He look older?
Not old, regal. Has a beard. No complaints from Me! I call Him Zeus when we do Greek stuff. If your Olympics were the Greek Olympics, the athletes would all be naked. More fun.
So You don't mind when hippies run around naked?
Not really. Many have great bodies, for humans.
But You want us to be modest.
Yes. Too bad your minds wander. Greek's minds wandered, too.
Gods don't need others to test Them? They aren't tempted by other naked people around Them?
Mostly right. Your mind is very limited...human. I understand more than you could comprehend.
I had a dream of my sister in a car accident? Is it real?
Won't say. Pray for her. Pray for all.
What if it's her time?
No such thing.

So praying helps?
Yes!
So is there a difference in me asking right now to watch over my sister through prayers?
Prayers change resonance and create new ripples.
That must be why prayers to even the wrong Gods help?
Works better with Us.
So it wasn't my brother David's time?
You helped. Pushed time. (Gave him extra time by praying for him) *Can only push so far.*

I had a vision that my brother was killed in an accident. The diesel he drove went off a hill or cliff and made a huge mess. I had overwhelming chills all over my body. It was the first vision I had ever had, so I wasn't sure if it was real. I tried to tell my brother but he had already left. I prayed all night for him and he returned the next day from his trip. I assumed he was either OK now or that I was just seeing things. Later I found out he had a few close calls.

Then one night I got the call at 3 am that he had crashed his diesel, had gone off an overpass and was crushed by his cargo. I knew then to pay attention to visions!

Who are the Hindu Gods?
False. Greek, too. Satan's confusion. Michael (Adam) *knew.*
Did they just start thinking others were overt?
Complicated. "God of waste" be intelligence. Satan's pagan Gods are made up. Over time the cult was deified. "Cult of Ryxi."
Me?
Yes! Be careful to stay righteous. You've been good lately!
Are you saying I'm changing resonance?
Yes. Red hair, would be a bad idea.
I would like to know more about the only good Queen that ever lived, and her records were destroyed by her enemies. It would be nice to learn from her what a good queen is like. Maybe even write a book about her?

Not yet. She's like me. I hug and spank.

Should we, "Judge not that ye be not judged."

Forget that. You will still be judged. Still must teach what you know, responsibility to teach. If you know a better way, you need to teach it.

What about "He who throws the first stone?"

Both guilty. Leave up to Us. Don't over think.

August 15, 2012

WASSUP?

What's all around?

Yes. So high, can't get over it. So low, can't get under it, . . . Peter, Paul and Mary. Hi Roy, Royal. (talking to Leroy)

Are you ever in a bad mood?

Was once. Spanked a lot all at once.

When?

The flood.

How many?

Over a billion.

They were all that evil?

Yes. Soon again. Fire, warming effects, famines, war, floods.

Over how long?

A decade.

Starting 2012?

More or less. 7, you like numbers.

Is 7 relevant?

Not much. I have short hair. Afro. Black.

Are you trying to shock Dad?

I never "try."

What gets you in a bad mood?

Kids not living to their potentials.

My son plays video games. I get annoyed. Seems he could be developing himself more and not waste so much time.

Improves reaction time.

Hey Leo, what's up?

Why are You in a good mood today?

Went surfing.

With Jesus?

He couldn't keep up with Me.

Really?

Yes.

Just the two of You?

Others, angels.

I like corn on the cob with chocolate. Try it.

Hey, Leo, you go journey tonight to meet guards (exchange) from other dimension or universe?

How?

No problem. Go with a dragon. First time. Opening. Exchange soon. Looks like a mint sandwich. I know this one.

Ryxi, your herb mother Maxine will be home soon from her assignment.

How does one look like a mint?

Not easy. Go smell it on the mountain . . . (singing).

Are you a fan of Peter, Paul, and Mary?

Only Paul. Nice abs.

Leo dreamed of graham crackers full of drugs and dealers trying to steal them.

Make s'mores.

Does it mean someone is trying to steal something?

Too late, no virgins.

What are You wearing tonight?

Best you not know. Dad's throne is not topless, but I may be! Of coarse only for Dad.

Why are You so happy?

God needs no reason, God needs no rhyme. God can be happy any old time. (Singing.)

God bless!

I did. Hallelujah! I be praised.

This has been a good session! Lots of fun.

Don't get used to it.

We'll take You as You come.

Good. I'll be coming a lot. Sang "I'll be comin' round the mountain . . ."

You should see my skirt. Never mind . . .

Ryxi and Leroy travel tonight and report tomorrow. Use oregano to ground.

How many times has Ryxi been to another dimension before tonight?

Once.

Twiggy (our cat) came from another universe. She stays here. Will go with Torg when he resurrects. Heal feet first, long walk. Leo, small mint 1st.

About Australia . . .will I run into my Shaman from New Zealand?

Try. Tell him to wear clothes and a pink carnation.

I see T-shirts in skate and surf shops that say,"Jesus is a surfer." How do surfers know Jesus is a surfer?

During Meditation. They know him. No surf board. (He walks on water, duh!)

Leroy went out to another universe and the next day reported back. They told him, "We are here to learn, lock, guard and protect."

August 16, 2012

Leroy is freaked out about You.

He's growing.

He's freaked out about your openness about sexuality. He blushed. But then so do we.....

He has issues with sex.

Chapter 18
Evil

August 18, 2012

I was feeling overwhelmed thinking that every time I put out a fire another one shows up. I realized that is how it will always be since She has said there is always evil everywhere and it makes us become Gods. A business I had started was being shut down by the government. I was excited to finally have money to create all our projects. Now I was discouraged.

Evil is eternal!

Yes, lots of you are evil.

I'm tired of fighting evil. It's an eternal reality!

Yes. The kingdoms are beautiful and peaceful. Gods and angels are worth the struggle.

Create a Petition. If Lot and his family would have stood up, Sodom would not have been destroyed.

I think the Social Security system is a Ponzi scheme.

It is! Not meant to be, but true. Never a good idea. Takes responsibility from the families. Hooterites, LDS, Inuit, Lapp, Yuimaru, and many others take responsibility and succeed.

Should families take care of the poor and disabled or should the country?

Families should take care of each other. Social Security tears families apart. Was supposed to pay families to stay together. Evil took over. Now it doesn't. Good people need to stand and be united.

Not my experience so far!

In the cities of Zion and Salem, evil lost completely. Many millions of good people stood. Needed to. Jealous, evil people try to twist numbers. If the judge says it was good, then people would believe.

When will there be justice?

Don't know. Agency and good people. Don't be bitter, Ryxi. Enoch's people are here and more are coming. The cavalry is coming!

Was the business legal?

Is legal.

Your law?

Yes and your Constitution. This is a very evil time. Worse than Noah's. That's why you are here.

Can't be worse than the medieval days.

Much worse!

Torture?

There is torture. There are more slaves than ever. Health is getting worse.

Are plagues coming that will kill many?

Heart disease, Cancer, diabetes, new germs; it's all crashing!

I don't feel like I've stood enough . . .

You have. This is a small ripple on the wave of this time.

I need to pray the judge will be honest.

And that good people will not give up. Needs to be proven legal. Government regulating business is unconstitutional.

We need to challenge the government! I don't have time! Where do I put my time?

This is a quick and easy fix. Welcome back Ryxi! My hair is red.

What is the quick and easy fix?

I got your back; you are now on my team! Petition to uphold the Constitution. Pursuit of happiness.

Can I have the exact words please?

The Constitution of the united States of America guarantees the right to the pursuit of happiness. This action by the government infringes on that right. We urge the judge to rule on the constitutional aspect of this action. Sign.

DON'T LET EVIL WIN!

I just don't know how to fight evil, I feel overwhelmed.

Me too. There is too much beauty to let him have it.

Torg went to the Philippines for three weeks to meet with local government and city officials to propose a waste handling system. If he gets the contract we would be well funded. This would be a good time to test my abilities to connect with Her in the trees without Torg.

August 30, 2012 (In the trees.)

I've been really tired, achy and crabby lately . . . don't feel much like a "servant of God" right now.

Must stay in hope and faith. Don't let the mountains (trials) fool you. My people have much time to waste during down time.

You need to rest now and stop worrying instead of resting! I am taking care of you, now believe it!! You are wasting your time on worry!

STOP, LOOK, AND LISTEN, We are the mighty Gods! (and old cheer-leading chant.)

We need $500 this weekend.

No problem! Sit tight, don't let the bed bugs bite. Hang in there. Your time is coming.

I received $500 the next day, by the way.

September 1, 2012

In the trees, Torg is still in the Philippines. I still just want to sit in the house. No people.

I'm still so tired!

I know. Comes with the territory. You have much responsibility and you're not taking care of yourself. Eat more greens. Sing, dance. Do only what you love. Sit in the sweat lodge tonight and pray. You are passing a test right now.

Suddenly it became windy. I asked the wind if it had a message. It said, "Hello, go with the flow."

What test?

Can't say. Shoulders and uncertainty. Much responsibility.

Breathe more for concentration. Money coming. No worries!

She sang: *WE ARE A HAPPY FAMILY!*

September 2, 2012. (Trees.)

Last night I did what she suggested and prayed in the sweat lodge; I got a vision I was in a cocoon. I interpreted it that I am in a cocoon stage. That explains why I've just wanted to be by myself. I usually like being around people but have been really irritated with housemates and quick to set boundaries to the point of being a bitch. Or at least I feel like it.

Hi, my daughter!

I feel like a bitch lately!

You are.

Why? Logically trying hard not to be. Would like to be warm, fun, and friendly?

Tough calling. Your job to be a bitch.

What?

Your calling is to balance and make things right. Sometimes balance is a bitch.

I thought You told me a while back diplomacy is better?

It is. You're tired, lonely, resting, irritable, and have hormonal issues. You're in retreat. People are irritating you because they are in your space and you need to be alone.

I'm tired of people in general.

Yes, Me too! Brats! Take a break for a while. You'll get over it!

I have sinus trouble again. What do I take?

236

Plantain / plantago? Pine? Rest! Drink water.

September 3, 2012 (with Leroy)

How is Torg?

With Pongust.

Leroy got the impression Torg is sick?

Yes. Love you. Owl twister.

Hi, Ryxi. You OK now?

Yes. Feeling better now. It's been a rough few months between falling down stairs and breaking my foot, my business being hijacked by the government , and feeling isolated.

Passed test.

What test?

Mind Me, rest.

What other tests?

Cocoon. Stand on your own two feet. No crutch!

How could Lot have saved more people?

By speaking up. Standing in his power. Warning others is a hard thing to do, standing against others. Scary. Worth it to save people.

Did he know?

He knew enough.

Were there other righteous people?

Could have been.

Sept. 14, 2012

How is Torg?

Sick. Misses home. To him feels like prickly there. Misses Twiggy. You, too. Feels honored there.

Are they buying it?

Close. Still concerned about viability. Unproven. Worried if it fails, there will be a big public out roar and they will lose face.

What is Your take on the Knights Templar?

237

Started out good. Like lots of groups, got twisted. Fell to evil. When groups twist too much we don't protect them anymore. Agency for all. Evil King win some. Went north to Switzerland and others. Did much good for a time.

What about the killing of pagans?

Necessary, but could have found other ways. Needed artifacts. Gold only to get temples and land. NOT for power. We warned. Didn't listen. Could have done much more other ways.

They murdered all those people?!

Not Our call. Acted on their own. Could have negotiated. Could have directed army. Told to leave, get city back. No need to kill.

People in Syria, wiped out?

Plague.

People keep coming into my life who have mental or substance abuse problems. Many of my friends believe the people who show up in our lives are mirrors of ourselves. I've also heard that healers will attract people who need healing.

Why do I attract alcoholics and mentally ill people?

Maybe resonance?

I'm mentally ill?

You have compassion, you give them hope.

I remembered I am also a Pisces. Many books say their best trait is compassion, but need to keep it in check.

September 21, 2012 (Torg is back.)

Torg and I dance. Save from elements. My and Torg's time. Didn't drown - "lots of water". Keep up good work.

While Torg was in the Philippines there happened to be several tsunamis and earthquakes in the area! And no where else. I wondered if it was because he was there.

Will he get a contract?

Yes, many. Small ripples, huge tsunamis. Many small actions make huge . . .

Torg was on television, TV action?

You cute. Many people saw. Beard makes you look smarter. Old wise man, St. Nick.

Torg: I thought I would go brown again.

No.

You change Your hair color?

I am God! Stay tuned to this channel!

Torg: My feet didn't hurt while I was there. Now they hurt again. Why can't they heal now?

Not yet, wait for the solstice.

Torg: Can I have a favor? Not hurt while in Australia?

Done. You owe Me.

Leroy came by and told me he had a vision of me in silvery shiny armor and heard the words, "Ryxi Bitch." Said he was going to start calling me Ryxi "B" like rappers call themselves! I still feel weird about the "B" word. My guides and the elements were calling me a bitch even before Mom did. I guess I need to remember it's a cool thing,

Leroy calls me Ryxi "B" for bitch.

Lion's not lying. (Leroy goes by Leo.)

Leroy: Who is G'nesh?

G'nesh is not an elephant. Metaphoric. Essence. Fire element. Teach in India. Not a GOD!

Healer?

Yes. All are metaphoric. Brahma, Rama, Rahib. Some are spirits.

Does it bother You that people call them Gods?

239

*No. They are Godlike. Powerful, wise. Arabs call
them djinns or genies. We call them trolls or dragons. G'nesh
is a fat fire troll.*

It made sense to me. I thought of another fire troll, Pele of Hawaii.
She was deified by the Hawaiians.
Leo also had a disturbing dream of a nuclear war coming.

Was Leroy's dream of war true?

Of course. Always has been war on the continent.

Nuclear?

*Yes. Terrorist could, hope not. But likely. They follow
false gods. Nuclear war has already been here with tests in
Three Mile Island against your own people.*

Everything the tree said is true?

*Yes, in context. Mohammed mostly was good, jihad
not.*

Did Mohammed misquote You?

*Yes. Infidels won't live. Don't need to be killed, they
just won't survive to the next level of evolution.*

The foot I broke has been itching a lot!

Healing.

I haven't been working on the book.

Lazy bitch. Just teach!

I am God. You are Me. Torg is Santa, hear him roar!

What about other Hindu Gods.

Hindu Gods are all elements.

Egyptian?

*Yes, Celtic too. Many. Norse not: Thor, Odin, Wodan,
Freya were real people, their real ancestors.*

Were they good or bad?

Conquered to unify.

Aren't we supposed to be sovereign nations?

*If don't fight. Genghis Khan was the greatest
peacemaker in history. No slaves.*

Is Krishna an element?

No, Krishna is Jesus, they have forgotten "Hari" is not a name. It means "Holy". Krishna means God....Jesus.

She must mean the word Krishna means God, but the man they worship named Krishna lived at a different time than Jesus, if I remember right.

Do You always show up to Leo looking like me?

Yes.

Leo says Mom appears to him in my image.

Torg, you study runes and acupuncture.

Thanks for the $3000! (I told her we need exactly $3,000 to go to Australia and the next day it showed up!)

You're welcome!

Did you do that?

Yes, should have asked for $5000.

Is the guard at Safe Haven scaring people? There are rumors of skinwalkers.

Yes! He and his friends. Things slamming, whispers, noises, lights, dragon moves. Funny. Scare burglars.

Locals say it is haunted.

He tries to scare people.

Before solstice 2012, what is most important?

Australia, Missouri, Giza, Machu Picchu.

After?

Baikal. no stress. Be a hermit, retreat for Ryxi.
Pongust talked to Leroy and Ryxi from Philippines. Torg made a new node in the Philippines. There are two nodes in Australia; one by Sydney.

Pongust was in Torg's pocket and we were able to talk to him from the Philippines! Makes sense, we've been talking to him from Stonehenge for years!!!

September 23, 2012

Talk to your New Zealand Shaman. Tell him you are coming.

I went into vision and met a Shaman with a white feather headdress from New Zealand a few years back and was told he was in charge of a node there. He was asking the universe how to take care of his node and I showed up! Hah, I'm sure he wasn't expecting a blonde white woman!

Go to sleep to talk to him?

You did before. You have the skills.

I went dancing and bought a taco on Sunday. Sorry for not obeying the sabbath?

Throne not topless. Jesus had to eat.

Knights Templar?

Both good and evil. All my kids are both.

What % are we?

90% this month.

September 26, 2012

Your lives are entertainment.

I have two friends who say they went to a Temple and had a sexual experience.

Yes, ecstasy. Should be normal.

One thought she had a sexual experience from Heavenly Father.

Ecstasy is confusing. Very emotional. Great passions overlap. People get confused.

Sometimes my body starts circling. What's happening?

Connecting to higher, Me.

Sometimes I'm at the computer and suddenly my upper body starts circling.

Looking for extra info. Antenna-like hair.

September 30, 2012

Happy Sabbath!

Leroy is on his way.

Hurry, let's talk about Sex! (Because He blushes)

Are you talking about this because Leroy is not here, since he went to get his notebook?

Yes. Innocence.

You tell us we can't have sex.

I'm a doer, not a talker. Celestial sex is always great. Like pizza is always great!

I had a dream about seeing a Tyrannosaurus Rex from far away. It was coming closer and closer until it came to my door. I felt he was trying to intimidate me. I was trying to show no fear. What was it about?

Teaching you to stand for rights. Hopefully it will stick.

Stick?

Ripples . . .

So because I'm a grid tied power walker, everything I do (not think) creates powerful movement of energy.

You're not a dumb blonde! Thoughts help, too. If you control your anger, it has a bigger effect.

So if I'm angry or sad and I dance anyway, it effects positively even though I am feeling negative emotions?

You are definitely NOT blonde. Talk about how flow works and I will not give $. Want cheese with that whine? Yet you deserve it!

I think she means that even though I whine about all our challenges, She still sends money.

When the continents broke up, did you need my help then?

No humans at that time. Ryxi Shaman Maori.

What's Dad's favorite color for your hair?

Blue.

When we are in Australia, should we stay on peoples' couches, http://www.couchsurf.org?

People's homes are more friendly. Go with the flow, ask God. That would be me. ASK GOD!

Where is my whale now?

Hawaii.

Why? Usually in Alaska now.

Came three weeks ago to do Shaman work. Will go back North to feed soon. She likes to be in Maui.

Leroy asked, Giza next?

Australia, Giza, Machu Picchu. Then Easter Island.

Oct. 3, 2012

Speaking with the elements:

How is the Grid?

Good. You have fun.

You've said before that Enoch and Moses didn't know you but they commanded you. Do I ever command you?

Not enough. Spank us! Women need to command more. Not want to command. Not their nature usually.

Both men and women dragons danced.

Why don't we hear about powerful females during Moses and Enoch's time? We just hear about "the patriarchs". What about the "matriarchs?"

They danced and sang.

Why are they not on the list of important women?

Dancing and singing are powerful for the Grid. The first five powerful Shamans happened to be men. Mary, Eve, Zippora (Moses' wife), Sariah, all strong. They were Shamans to keep balance. Mary told Jesus what to do.

Oct. 8, 2012

We had to deal with an issue at Safe Haven where we needed to stand our ground against a manipulator. We asked a friend who teaches what he calls "Verbal Judo" to help us stand in our power.

What do you think about our classes?

Verbal judo is a good start in a long study of diplomacy. You be careful to use skills to change people, not agency. If use to manipulate, you ask Me. Hitler and Mussolini used the same method. Educate, learn well, follow spirit. You have just started a long study.

What should we do next in importance?

Bury a library with information to fix society, clothes, food. Tubes at Safe Haven.

Book?

Write everything in one book. Magic and all. Only time to write one book before solstice.

A friend called and said she was guided to pick up white rocks and wondered if they were seer stones. She had heard of a prophecy that everyone would have their own seer stones during the Millennium. I told her I would ask Mom.

Are Dina's stones seer stones?

Be careful. A seer stone could see too much. Can be Dangerous.

Should she still use it?

Ask me. Knowledge can be dangerous. There's no going back. This bard helps. Responsible for what you learn.

All knowledge is dangerous.

You are frustrated deeply with new knowledge because your old life is gone: family, friends, sons.

She was talking about how my family and some friends or people at church think I'm really "out there" and are uncomfortable around me. I feel ostracized sometimes and complain about it. But at the same time I have many friends who are on my same page.

Is it all worth it to You?

You. Me , and Gaia. You want to be Gods? Most of the time, bigger boxes are more vulnerable.

Chapter 19
Uluru &
The Rainbow Serpent

It was October and only two more months until the Winter Solstice! We still had so much to do. For the past few years we were told we were supposed to go to a lot more nodes around the world. So far we had only gone to two: Mesa Verde and Pele in Hawaii. I was worried we weren't doing enough. But how could we get to all these places with no funds? Time was running out. I had received enough for the trip to Australia, but was concerned we would not be able to get to the rest in time. Except for Missouri, which was within driving distance.

Uluru

I decided to first fly into Brisbane and check out the rainforest and beach of Australia. Beaches are my favorite! We would then fly to Uluru, a huge red rock in the middle of the outback I saw on the internet, that the indigenous people reverenced. I figured if they reverenced it, it must be an energy center. The plan was to then drive to Alice Springs from Uluru.

Even though our target was Alice Springs, I didn't want to go to Australia without visiting the rain forest and the beach! Besides, we had a friend who helped us start Safe Haven and had moved to Australia. Karen and her daughter were living at an intentional community called Billen Cliff, how convenient! I wanted to visit their community to compare notes and become connected to others around the world to work together.

Her new husband picked us up at the train station and was dressed like an Aussie with a leather cowboy hat and all. I felt like Crocodile Dundee was picking us up! It was fun hearing him speak the Aussie lingo as he drove us for an hour to their home.

I was fascinated by koala and kangaroo crossing signs as we followed the winding jungle roads through the rain forest. I was thrilled to see a wallaby hopping across the road as we drove up to their home.

Our friend Karen and her daughter Joy were happy to see us and we stayed with them in their off-grid home. Their community was like ours but had been around for a lot longer. It was interesting to see how they catch their rainwater and use solar panels for electricity. They too were healers, earth lovers, and indigenous empowerment supporters. We were honored to bless and create a medicine wheel for one of the families, and to teach Matrix, as well as do a male/female ceremony for the community.

One of the community members heard we were going to Uluru and told us of an aboriginal man named Uncle Bob, a friend of hers who had created a documentary about Uluru called Kanyini. We watched it and became very excited to go there and meet Bob.

Uncle Bob is what the Australians call a creamy. They were half white and half aboriginal kids who were taken away from their moms by the Australian government to "breed the black out of them." The government only recently apologized. In the movie he shares his story of living in concentration camps as a boy. You can see it on You Tube under Kanyini, Bob Randall. Very sad.

Oct. 29, 2012

At the kitchen table listening to the rainforest frogs sing with Torg and Karen:

My hair is rainbow. I went bald last millennium. Dad didn't like it.

Sinead O'Connor is bald and beautiful.

I sing better.

Uluru is part of the Ley line system. Like a chakra. Not a node or nexus.

Similar to Sedona?

Yes.

How many like them are there?

14 on the whole earth. A spirit connection, like the movie Kanyini describes. Like Sedona, a place of healing. You will find others when needed. Antarctica is stable.

Wasn't there anything you could have done about all indigenous people?

Your ancestors were oppressed too. All people through history were. Some worse than others. Transition (death) not as bad as you think. They only came home.

Yes, we don't remember our home. . .

If you did, you would come home now.

Tomorrow, get in the ocean and talk to the squid.

What should I say?

Hi, squid. (Squid that helped with Madagascar)

Big Water: *I'll roll you over in my waves like in Hawaii and get sand in your suit.* (My swim suit was filled with sand and looked like I had a diaper on. Every time I got up he would send another wave and push me down and I'd roll over in his waves.)

Yea, well I got a new suit so that won't happen again.

Not as much fun.

Uncle Bob?

Great talk.

Is he a Shaman?

He doesn't think he is. He is into it. He should accept his calling. You honor him, teach him his calling.

Is he going to be an indigenous Shaman?

No. He can't accept himself as a real Aborigine. You help them, honor them. They listen.

What can they do?

Teach people about their culture. Teach about the government. If enough people are talking, things could change. Keep doing what you are doing and more. Teach new economics. You go to Uluru, make a difference. You know the Hopi. They are taught to want the wrong things. Ripples.

How? Not much time.

You're the Ryxi!

How did we help the Hopi?

They trust whites more. Open to more change. Teach spirituality. No more secrets. They need to clean up their mess! You help.

Why isn't anyone else helping, we can only do so much?

They are. Local hippies, Buddhists, this community, etc. You're in Australia. Everything is magic!

I swam in the ocean and tuned into the Shaman squid who

helped us with Madagascar. I guess Mom wanted me to connect with her because we were closer to Antarctica. I just got the sounds, "Blablum, Blablum...."

Our friends took us to the airport and on the way I got sick and threw up in the car as we were arriving at the airport! I didn't have time to clean it up! They had gotten the flu while we were there from her daughter's school and we had hoped we would escape it. Torg started getting sick on the plane.

When we got to the airport at Uluru, there were no rental cars left. It wasn't a regular town, but a resort base on tourism. Tourist came to see the rock and go on outback adventures. I had planned on driving a rental car to Alice Springs. I was really upset. We were stuck at the airport with no car and both of us were sick.

I had spoken to Uncle Bob on the phone and he had invited us to stay with the aborigines at their community next to Uluru. I was supposed to call his son Johnny to let us in, but he was not answering the phone.

I said a prayer and soon felt prompted to talk to an Aussie tour guide and ask him about his leather hat with pins that looked like boy scout badges stuck all over it. He introduced me to another tour guide, Leroy, who knew where Johnny lived and was willing to take us there on his huge tour bus! Johnny worked with him and had a place next door to him in the campground where the employees stayed.

We finally connected with Johnny and he informed us they were having Men's ceremony at the community and I would not be energetically safe there. We had a good conversation and heard about his experience growing up in the concentration camps as the son of a creamy Mom and Dad. I was so sad. We decided it would be best to stay at the campground instead of at the community. I didn't want to be in danger, and I didn't want the abos to feel uncomfortable.

Our new friend Leroy was so nice and gave us a ride to our campground spot. The campground was amazing! The sand was dark red and reminded me of Southern Utah. The trees were an interesting shape I hadn't seen before and parrots were landing near our tent. We saw signs warning people of Dingos. I still didn't know how we were going to get to Alice Springs.

We caught a very expensive bus tour from the camp ground out to Uluru to do our ceremony. The Tour guide's name was Pippin and looked like an elf. He told us a lot about the aborigines and how they were fighting to keep Uluru sacred. I could tell he reverenced the rock and her people very much. He also talked about how they were trying to grow their own food. It hit me that if I had been able to rent a car I wouldn't have met the locals and gotten the real story of Uluru.

We stopped to see the sunset and our other tour guide friend Leroy was also there with his "mob". (That's what they call a group.) We performed our Male/Female ceremony as the sun went down.

We went back out the next day to spend more time with Uluru. I tuned into her and she said, "My womb is your womb, my rock is your rock, my heart is your heart." We performed another Male/Female ceremony in a deep alcove.

November 1, 2012

Again, we were at the campground. The night before, we went to the resort club and Torg played the drums for the entertainer who loved him. I danced with the rest of the tourists. By the time we got back to the campground the wind had started to blow hard. My tent was blowing sideways and almost flat. There was an eerie feeling about the wind. Something was different. Even other campers noticed and commented that it was freaky. Some had planned to sleep under the stars and chose to sleep in a tent, it was so weird.

By the time we talked to Mom I was upset and on the verge of tears because the buses to Alice Springs were $300 for each of us and there were still no rental cars available! Would we have to hitch hike? I thought I had messed up. We had been working on, and communicated with, Alice Springs for years! Our whole trip was based on Alice Springs and now we weren't going to make it!

Everything in Australia was double the price in the US and I had been stressed about money the whole time there. It seemed all we did was teach Matrix, visit a community and swim in the ocean. I couldn't see that we had actually accomplished anything substantial worth spending so much money for the trip. By the time I talked to Mom I was a mess!

Love you.

I'm a whiner.

It's your nature.

Not all my life.

Much at times. Very pleased. No more games.

So You are admitting this is a game?

Bad word. We will have someone else go to Alice Springs and do it. There are many other choices. You were the best. No worries!

I thought you guys were Gods? You can make miracles!

The big picture.

So we came all this way just to do Matrix? Aren't You supposed to be supporting us? If You ask us to do a job, don't You help us do it?

It's hard for you to see. I Am. If it is too hard, we can release you. Your job is the Grid and service. You've done enough. Your place is secure in the kingdom.

It's not just about a place in the kingdom. I wanted to accomplish our goal of aligning and strengthening the node. Torg was supposed to meet the guards there too. I feel like a failure. (At that point I was sobbing)

If you are stressed, don't go. You did more here at Uluru than you would have done at Alice Springs.

How? I'm not a healer like Torg.

You are a healer.

I'm an angry bitch.

You are a bitch and a healer. You talk to people, make connections, wake people up.

I was supposed to come to Alice Springs and do node work. I didn't do my job: the thing I'm good at.

You did do node work here. Woke up many nodes. The nodes are much stronger now.

How can it be stronger when I'm angry and frustrated?

Would be better if you weren't, but you are who you are. You can't make a splash like that from a plane. To wake things up, you need to walk, dance, drive, hear them to wake them up. You feel the wind? That's things waking up all over.

I can't feel or see the impact. I just take your word for it. Torg can feel the difference and knows it's working. Torg thinks I need constant reinforcement so I know what I'm doing is working.

You do. My bad. I will from now on. You woke up more than all of them here in years.

Sorry for being a baby, but I can't see the results.

I should know better. I am crying for you. If you had had money to go to Alice, it wouldn't have worked as well. I need to trust your team as well. You "fullness of times" kids are not what I am used to. I will try harder to be part of the team, not just give orders.

Huh?

I was shocked. It was like she suddenly had an epiphany. Her attitude had turned on a dime and now she was apologizing? This was a first.

You don't feel things, Torg doesn't see things. You are grown, not like the Israelites who needed to be told what to do. We won't treat you like babies. Can you handle the truth? Torg, it's like when you do scout training, where you set up a situation where kids need to learn something, and then they listen and do. They learn through the exercise.

I still don't see what I do.

You woke up the entire coast and all the entities are saying, "What the hell?"

No wonder the winds were freaky. The winds blew hard last night. I had a strong urge to sleep on the ground outside my tent and feel the wind all night. It was like it was calling me and healing me. Besides that, my tent was blowing sideways.

It's part of the splash, the wind is blowing. You came, you did, changes are happening now. You can go kiss the rock if you want!

You said You need to trust my team. What team?

You, Me, the elements. I will ask you your opinion from now on. I am not used to this dispensation. Even the guard is new to Me. Always in history before, I commanded and things happened.

Are You saying that the guards, elements, and I decided it was best that I go here instead of Alice Springs?

I need to remember My own history. It's been a long time. Billions of years, no big deal. I promise not again, forgive Me.

What am I forgiving You of?

Not telling the whole story.

Do You mean generally the goal was to wake up all of Australia, not just the node? You said to go to Alice Springs, but my team and I decided that I could do other things that would work better?

Yes. Like California, Guatemala, Hawaii, Sanpete, Hopi, etc. You are noisy, and wake things. This is not Torg's skill. Torg talks to the guards. Ryxi is a walker, Torg a talker. Ryxi wakes up, Torg tells them why. Ryxi is an alarm clock. I told Enoch, he told the elements, things happened. Things are different now.

Because the elements are growing, too?

Yes, end game. So relax, kiss her rock, etc.

So did You tell us the truth and wondered if we could handle it?

Good logic. You would make a good attorney. Do you like the temperature? (The locals told us it was unusually cool for this time of year. It was usually in the 100's) *Your friends* (the elements) *say you are welcome. You are in the flow and have no flow. Keep a list in your pocket of the miracles.*

Nice Ryxi.

Thanks for hitting me with a lightning bolt.

Torg, you spank her; cheaper than a lightning bolt!

List of Miracles

1. Stupa. Free of charge. Woke crystals up. That's why Torg is off balance. Portal is stronger. (We went to a Buddhist stupa dedication near Billen at a place called crystal

254

palace that sells amazing crystals. Torg was off balance cause they were so strong.)

2. Nimbin now more of sovereign people. You made connections like you made while in Tikal, Guatemala. You got the call from Hopi through your connections with the Mayans. (We visited Nimbin, the famous Hippie center of Australia.)

3. Leroy very good friend now. Will help move art around the world. (One of the tour guides who's wife sells aboriginal art. We connected them to art distributions for aboriginali economic development.)

4. Uncle Bob. Stronger now you talked to him. He's commonplace. Has an energetic connection to Alice Springs.

5. Johnny contact, knows aboriginal perspective, complex person. Conference for indigenous peoples in Vancouver when you have money. He has feet in both worlds. Because he is a creamy, both white and Aboriginal. Understands more than Bob. You will work with him in the future. (Johnny is Uncle Bob's son. I had an idea to create an indigenous conference where tribes from around the world come together to share information on being empowered. If they work together and form an alliance, they are more powerful.)

6. Pip. Followup with him. He had a tear in his eye. Likes you. (Another tour guide that spent time with us before we left and took us to the airport. I shared with him about the elements and he has had experiences with the elements too. I gave him a copy of *To Dance With Elementals*. The elements call him an elf. He looks like one too. He looked like he was going to cry when we left. I guess he doesn't meet very many people who hear the elements. Must be lonely)

7. Uluru. Torg's dream is right, came from orbit. Rock was thrown, not made by a volcano. Blew out from South America and came back. Happened during dinosaur even when the big rock hit South America. (When the dinosaurs became extinct.)

8. Billen Cliff. Wheels all over.

9. Beach: played with Big Water. Contact, no box. Squid with whales heard you. The whole area is still twanging. They felt a loud psychic sending. Nice Ryxi. Blub blub, the squid you heard speaks for all.

10. Rain forest: Tried to keep you there longer, felt good. Full of life. By hurting Torg's knee! Not My idea. Not My idea. Trees too. (When Donny took us to the rain forest we got lost and Torg hurt his knee. There was also a Python on the trail with us. Torg was thrilled.)

11. Music: Torg's drumming made a splash. Made people happy, brought positive energy. Torg helped, impressed, ripples. Many people took pictures, went home. Positive energy saved marriages. A man came to get away from wife, was planning to divorce her, went back to her. Torg is healthier because of it. (We went dancing at a club and Torg picked up the drum and played with the singer. It was fun!)

12. Rock (Uluru) is a She. A bad entity tried to stop you. Sent lot's of flies, Torg's feet hurt, tempted to leave the trail and get a ticket from a ranger.

13. Karen caring, good friend. Nice like Ryxi. Both Goddesses. Serve.

What makes Uluru so special?

Because she is alone.

Alone in the desert?

Partly. An energy center because the only rock. Very dense. Uluru is strong, like Pongust. The rock landed on a ley line. Is almost a giant. Although is stronger than a giant. Maybe more alive.

November 3, 2012
Uluru campground

Uluru is a she. You did well. Bad tried to stop You.

Was the message I got real?

Yes. She liked the Battle Hymn. (I sang the battle hymn of the republic to her when we did our usual dedication to Christ ceremony)

Does she know Jesus?

She will. Likes you Jesus people better than earlier Jesus people. Priests corrupted the people: new ways, new food, new money, new laws.

So I'm going to put my faith in You sending money flow, and buy knockers! (Some knocking sticks I had my eye on that are used as an instrument and painted by a local aboriginal artist with signature aboriginal art dots. Was worried about spending money)

You have nice knockers. Torg do dots. I will try dots on Dad.

Brand him?

Can't; God bod.

Good job Torg, give Chris (a random Shaman we met who was from Alice Springs and was heading back there) *the Lake Baikal rock.*

Who is the backup Shaman for Alice Springs?

Chris is.

What is my job if I don't have to be there?

Power up Grid and fight evil.

But I can do that from anywhere.

No, best if close. Nodes are usually best, but not this time.

Is the node powered up then because we came?

Very much, way to listen, yay team!

Should we have gone to the Olgas (mountains close by)?

No, they are male. My millennium is female. Alice Springs is the node and center, but Uluru is the heart and soul.

The rest of the trip I decided to have faith and have fun. I splurged and bought the musical knocking sticks with aboriginal art by a local artist. I decided to 'Have Faith" and let Her figure out how to pay for it. I imagined every worse case scenario and laughing about it so I wouldn't be stressed. This was a turning point for me that

I hope will last a lifetime. I'm sure Torg appreciated my attitude adjustment!

Karen's husband Donny (Crocodile DON-dee) picked us up again at the train station for our last few days in Australia until our flight would leave from Brisbane back to the states.

November 3, 2012
Back in Billen Cliff with Karen.

Hi kids. Dad didn't like the dots.
Karen: What is the legend of the rainbow serpent? Aboriginals are waiting for the rainbow serpent to bring peace.

Many people come and join together. Many people are the Rainbow Serpent. You see other truths coming together? Ryxi's calling. "Rainbow" meaning "races"? Tied in one (Rainbow tied in a knot). The New Age is about joining old/new, East/West, mysteries no more.

She started dancing the infinity sign.

I like "I'd Like To Teach the World to Sing." I prefer nut trees to apple trees. Almond Joys!
The legend says it will start from the East?

Bigger box.
What does "East" mean?

Asia. All natives have Rainbow prophesies. Yes, even Vikings. After a storm comes a rainbow. Noah story type for last days after End Game.
So the races clashed in the beginning, then will come together. Did mixed kids sign up for the change over?

Some. Yours did. (Karen's daughter is mixed too.)

Yes. Tell the Uncle Bob story. How to teach people love? Long story. Song (Everything is beautiful).
When will the Rainbow Serpent come?

Came, saw, conquering.
When did it start?

During the renaissance. Picking up speed. Like a Phoenix rising.

Suddenly someone new came in, it was Uluru! I was excited to talk to her.

Uluru: *Call me, just call me. LOVE YOU! You like the rock? Torg has baby Ulu rock.*(Torg felt he was being gifted a rock) *You treated my place well. Thank you for your rocks and ceremony, it tickled. Honor me, take my rock with you all over. Goddess Karen sleep with us tonight.* (Pongust and Uluru)

How do you feel about people walking on you? (Uluru is a tourist attraction where people climb to her top)

It's OK if I invite. Not many. You could, not Torg.

Because he is male?

As far as he knows.

Do you not want males on your rock?

Not often. You go on sacred mountain only if invited.

Will you tell us more about you?

Born of fire and earth baptism and renewal. I am the only one.

The only one like you in the whole wide world?

Yes.

Why are you sacred?

I'm a Big rock. More than just a big rock.

Great big rock?

Gaia needs a solid anchor.

How can we support you?

Go teach mysteries. Teach about The Rainbow Serpent and races coming together, "The joining."

Sounds like a movie.

Nodes are centers, nexuses are joinings. I'm a wow!

Did you know we were coming?

Yes.

Who told you?

A Shaman.

Which one?

You.

Did I come visit you out of body?

Yes, nice body. Met your friend.
The New Zealand Shaman?
Yes.
Did we come together at the same time out of body?
Yes. More than a month ago. Sat on me.
Did Torg visit too?
No. Hi, Torg. Not on me.
How did you get here?
Big boom. This continent.
From here?
Here has moved there.
Put my rock in chocolate milk. Wow! (We had hot chocolate.)
What was the most powerful of what we did?
The male/female Ceremony. Black and white sand. Waiting a long time for this. Mom's time.

Our Australian friends were worried about a coal seam gas drilling project being implemented. Many were protesting against it, especially Donny and Karen. Karen was hoping the elements could go in and sabotage the drilling.

Karen: Are the elements helping with the coal seam gas mining problem?
Yes they are. Gaia is on it.
How can we help?
Dance, pray, love, sing builds up the grid. All help, helps. We need Grid power. Ryxi knows. We drain the Grid for special needs. When you help the Grid we have more power to use. Give chocolate.
Can the elements oxidize machines?
Yes, they're doing it. Magic dance. Ryxi worked on the power grid in Hawaii first. No, the nexus at Tikal Guatemala first, then the nexus at Hopi. Gaia chose these places for Ryxi to visit for the purpose of balance for the Grid. Gaia here:

How are you?

Gaia-ish. We like it when you tell Torg where to work on you next. I am the same. Next I want you to work on Cairo.

So you say, "My Australia hurts." That's how you determine where we go next?

Yes. "I can't feel my butt."

Where's your butt?

Antarctica.

Where is your heart?

Uluru.

Where is your head?

Pele.

So your heart feels better?

Yes, you're Maalox. Scratch my back.

Where is your back?

Tibet.

Are we going there?

No hurry. Maybe July 15, 2014. Dance, fart, etc.

How do we hug you best?

Go to caves, crevasses, deep places.

So the place where we did the ceremony in the cave was good?

YES! (Stuck.) How many dragons does it take to change a light bulb? They ARE light!

You gave us malachite. What's special about malachite?

It makes things believable. When you teach "no disharmony", students believe you. Use wisely Obiwon.

In the cave I felt Uluru asking me to leave a crystal. Why did you want the crystal?

It's not like me. Tasted good. Made me shiver. No . . . like you people say, sex! WOW! Shiver, what's the word? . . . Orgasm!

Who felt it?

People in the area. Nice package. You come back please? Karen, will you visit me and bring lots of different rocks?

Don't people bring you rocks?

People take, not give. I'm a giant. Not big giant like Julie in Hopi.

Are you happy in general?

You bring new wow.

Do the aborigines make you happy?

Of course, they pray and sing. They need to grow now.

How?

Healthy male/female relationships. The guard is a nice guy.

Karen: How can I share my talents, what are my talents?

Innocence. Accepting of people, people skills.

Sleep. Go to dream time. Good Ryxi, pat on butt.

November 8, 2012 - Home

Uluru: The wind came and woke me up!

Weren't you awake?

Not really awake. Wind awoke, came because you woke up others.

How are you more awake now?

I only Knew black shamans before now. I thought little of whites.

Now you are open to whites being good?

The big picture! I thought whites were evil.

Now?

White shamans are more advanced. Working on Gaia all over, not just this little piece. My job now is to wake up blacks to modern purpose, Gaia is not just one rock in the desert. They need to join modern Shamans of all races.

Will they listen?

Don't know. I haven't asked them to do anything new in a long time. Tell them to join the world Shamans.

Sounds dangerous. Indigenous shamans don't acknowledge or except white shamans.

You are strong and don't understand how strong you are. Don't know what you are doing. You woke us up.

I hope it was a good thing. Heavenly Mother tells us what to do.

She doesn't ask? You are stronger if you understand what you are doing.

Are there more rocks like you on the earth?

Five or so.

Did you know we were supposed to go to Alice Springs?

Not much going on there. Home of big earth guy. Boring. Not like me.

You're more exciting?

Very.

Any strong Shamans there?

I am strong. Not as strong as you.

We are still understanding our strength.

You need to wake up!

Do you let both males and females on you?

Yes.

Why not Torg?

Not his path. Pongust says I should trust you. Will you take care of my baby? (The rock she gifted Torg)

Torg: Yes I will.

Other rocks like me are: Pongust, size doesn't matter. Easter Island, Machu Picchu, Eastern Canada.

What makes you unique?

Not us, who's inside the rocks! It's a great honor for you to talk to me.

I thought Gods were humble. I guess they aren't.

Yes, know who they are, humble. We make fun of ourselves.

263

What do you do all day?

Watch.

Uncle Bob invited us to go stay at the aboriginal community, but his son John said it was not safe because they were doing a men's ceremony. Why was he so worried?

Bad karma for them if you got sick.

Was it real?

Yes. Would have made you very sick, maybe die. Same for female ceremonies, make males very sick.

Are the ceremonies necessary?

Were.

Still?

I don't know. My guard says it's the same all over Gaia. The Watusi and Hopi have them. Not everyone. My guess, not good any more. It was not good to go to the community at all, ever. You came at the perfect time.

Why?

Many synergize.

Must be a shift we didn't know about.

You did much.

We wouldn't have met Leroy, Pip, Bob any other time?

The way the universe works. One wrong ripple wrong.

Anything else we did for Uluru besides wake you up?

That's a WOW. I haven't talked to Pongust for a Long time.

Is Figi awake?

Yes. In the process of waking up from your chanting in the basement of the airport.

While we were on an 8 hour layover in Figi, I went down to the basement of the airport in a private place where I danced and chanted.

November 11, 2102

Mom: I am doing much. The busiest time of my life.

Why?

It's MY TIME! Setting things up is always the busiest.

This is new for You?

Yes. Rethinking, not easy. No more commands.

For us or all?

You and other big kids. So don't let Me down! Not appropriate to spank team members.

Yeah, no more spanks for me!

I can change my mind. I can tell Dad or have Torg do it!

Why do I have to work so hard at staying positive while doing my mission? Feels like I'm going against the current.

Look at history. Your calling is depressing.

Money's the issue that's making me depressed.

Other things would have happened If you had more money. Money would have stopped your calling.

I would have chosen other paths?

Yes.

Even now?

Some. Your eyes are now open. Very few happy rich people. *Very few rich prophets. Time for $$$. Let's not discuss money, you get frustrated.*

I need money to get the book published. Do I put everything in the book? There are so many subjects I don't know what category to place it in: magic, Christian, science, New Age, history, pagan, etc.

Yes, the right people will get it. How many people understand the scriptures? All who have written sacred works have these same feelings. Very few People know the Bible. I would say most people are . . . sheep. You are not writing for sheep. You are mainly writing for crystal kids. Kids' stories. Don't talk down to them. Talk to them like adults.

I guess if the earth is deceased, the new kids won't know the past and will start fresh with new scripture?

Including yours. Yes, that's what I mean. Most people are still people.

I read writings from the book of Enoch where he wrote that men wearing women's clothes and women wearing men's clothes is evil?

It was the Enochian society law, not our law. Homosexuality is not evil if one is truly homosexual and monogamous. Our call. Ask Enoch when he gets here.

Will we meet Enoch when he gets here?

Probably.

Will he read my book?

Yes.

November 18, 2012

What does false humble mean again? It sounds to me like it means pretending to be humble when you are not.

Can be both. 1-pretend to be humble. 2-Feeling without reality. Not having enough confidence. Stand in power you really earn. Admit to what you really know, what you can do, what you can be. No more or no less.

True ego is in line with God. Too much is a wild card, too little is useless. Ego is not bad if you are on the same page as Us. Ego gets things done. Ego can mean either too much confidence or being in alignment with God. Humble can mean either teachable or not enough confidence. Schematics always causes problems. You should be a lawyer.

I am barefoot. Danced infinity sign.

How many people will die during this cleanse of fire?

How many is a lot?

As many as the flood?

No.

So it won't be as bad?

Yes, millions not billions. Not all at once, it's a process.

Chapter 20

The Climax

November 21, 23, 2012

(Special "Dragon Dance" message from Prescott to Torg)

Torg went out of body to talk to Prescott on two nights. He knew he was going on a trip because he was craving chocolate so much. The subject was the coming dragon dance, of course. And many more were present than just Torg. Prescott told Torg that he should give the "Ulu" stone to Rider, after soaking it in Nolmi salt water. She was to take it with her to the Rainbow Gathering in Mexico, as a protective link to Uluru, the "Heart" of Gaia. This makes sense, being on the empathic female version of the dragon dance at the beginning of this Millennium!

Sean (a pre-mortal spirit who considers Rider to be his mom and Torg his dad) will go with Rider as well. Torg's dad, Earl (who is resurrected), will also go with her, since he considers Torg to be well protected and in a familiar place, and he feels a need to make amends for some probably bad advice he gave the two of them.

Prescott also told Torg to update and print his Will and Last Testament; which he was to give Ryxi. Torg gave the same message to Rider. Ryxi is not to open it unless he is gone more than five hours, at which time the connection between Torg and his spiritual/physical body will be too tenuous for him to find his way back. Apparently there is a danger that some of the Humanized dragons will become "light lost" and not return to their corporeal places. He says there is about 300,000 humans in this condition, and many of them are in dire straits at present. Torg and Prescott feel this danger is very slight for either Rider or Torg, but still wanted the safeguard in order to cover all bases.

It also drove home the very serious nature of this last dance and the potential risk for the participants. Torg is to keep Pongust in his left pocket while he is traveling, and his onyx in the right pocket (and carry nothing else). He is to set up a special medicine wheel in the dirt under where he will be leaving his body and spirit. He is to smudge the area well, and fast for 24 hours prior to leaving at noon on the 21st of December. Just prior to leaving, he is to eat and drink very dark chocolate. Prescott gave Torg a list of the best six men and six women to be his guards. More of each would be OK, but he should try very hard to get at least six of each. The guard will perform whatever ceremony Ryxi chooses, and then do a sweat lodge while Torg was traveling (they need to stay within 50 yards while he is gone, but do not need to stay in the circle.)

Nov. 26, 2012

Prescott here.(orbiting)

What's up?

Me. Just here to watch.

Mom: *Love picture, looks like me!* (Book Cover)

The eye color?

Sometimes.

I did what you said...

Godly. Idea, Godly.

I would like to write songs and make a CD to go with the book.

Yes, lyrics like, "Mom speaks to me....
Mom speaks to me, can you hear her too?
I can hear her, you can too.
Ask and listen, receive.
More on Wednesday. Four verses and a chorus!
"Give said Mom to the little stream..."

Good ideas. Of coarse, you're God.

I AM, Wow me!

I play the Zithrax, an instument here. You will play the Zithrax.

What?

Like a violin you blow in. A didgeridoo plus violin.

Can we call it a Zithra instead? Zithrax sounds like anthrax.

Already one.

Is there a Zithrax here?

No.

When do you want us to go to Missouri?

ASAP. Drive. Watch while driving. Wake things up.

I went before to a Nemenhah Council gathering a few years ago, didn't I wake things up then?

Yes, wake more up. More important to get there fast than wake things up. Give Nolmi salt to the Mississippi river.

Our friend Marcin lives only 70 miles from the Node. He said he was inspired to start a community there to restart society.

Yes.

Will he need to restart society?

Near, good to be prepared. Good to have a mechanic with machines around.

Our friend Dex asked about getting a wife.

Pray to Mom about relationships, not Dad. No arm candy. Pretty face gets old and wrinkled. Not me! But you will! You're all on that same boat! (The three of us are single) *Big boat! Relationships are broad, many aspects.*

Dex left and went to the restroom and came back...

Feel better? Gods don't pee.

My gown is green.

Dex: I was told in a blessing to ask the Lord for advice on whom to marry.

No, you ask me. Otherwise good advice.

Torg, no more life of Job. You still have to work with Ryxi. Ha ha!

At least we have the same passions.

Yes, no sex. (We're all single)

You 3 are a pair of boring...no sex.

What about sex?

I have never....never mind.... Tried a clear gown. Buy a fig leaf. No need for a leaf. Try on Dad's.

We all rolled our eyes and Dex blushed. It seems She does this on purpose when new people show up to talk to Her. I think she gets a kick out of watching people blush!

Do you have anything important to tell us?

Tomorrow, I'm going to go see Dad.

What do we do about our business that the feds have taken over claiming was illegal?

You followed the law. I'll ask Dad after I'm done.

He's never talked with us....

Silent type. Your business was legal, make much moola. Make Love not money. Gods don't blush.

Will you tell us what Dad said tomorrow? Will Dad come and talk to us?

Try, he will be tired.

God's don't get tired!

Teasing. Make spirits not drink spirits.

Dex: Am I on the right track?

Yes, driving backward. Another train coming. Nice caboose. Engine needs work.

Ryxi: Is he doing his mission?

Some. Bring in my Millennium. Get new shoes. Long hard walk. Walk consciously, live, marry tomorrow, my time.

Anything more for Dex?

Shake rattle and roll. You're a God, go get a Goddess.

November 27, 2012

Slimming down for my picture for the book cover..

Try spandex to look slimmer. But guys will hit on you. You will be distracted. I sent you 22 year olds. You needed a boost.

There were a few 22-year olds who wanted to date me a while back. It was nice for my ego but weird! My eldest son was only a year younger! Torg had dated someone younger than his kids. It seemed there were all these people dating people way younger than them. I had another girlfriend who was dating and then married someone 16 years younger. I guess if guys have always done it?

Did you send them on purpose?

Maybe. You needed a bust... no, "boost". You have a bust. Opened the way for them. You like the way?!

Yeah, it did boost my ego. They're all 23 now.

Too old!

You told me I had to grow more.

Not much now, I send 23 year olds?

270

Can't you send some my age?! Did you have the report on the law for our business issue from Dad?

No. I was distracted. Clear gown worked!

Angels on deck for tonight. (Rider was scheduled to teach our Shamanism 101 class how to talk to their angels.)

Dec. 2, 2012

While in the trees I received a few scriptures and was told they were about Her. I thought she said she wasn't in the bible. I was confused.

Are you in the Bible at all?

Yes.

What scriptures are you in?

John 5:7-11 Take up bed. Not my law, Man's law. You look it up.

Zacharias 27: I was taken out. Most scriptures with empathy, family, healing are about me. I was taken out.

Jesus healed and was not taken out.

My skills!

So when He healed people, He was using Your priesthood or his?

Not that hard. You all use priesthood as you see fit.

You've said that priesthood is acting in God's name.

Yes. Empathic more in line with me, knowledge and law more with Father. Both are the same: the authority to act in Our name.

The original question was, when was Jesus acting for You?

When he cast out devils, more like Dad; when raising the dead, more like Me. You ever spank your kids then hug them? Which priesthood were you using?

I guess she means when we spank, it is Dad's priesthood we are using because we are disciplining for breaking the law? And when we give hugs we are nurturing like her? Seems unfair for the guys, they get the dirty work.

Did you find out from Dad about the law for my business issue?

271

Yes. You do no more. Let the courts do the work.
I'm confused, you said to read Revelations 5:4 when I asked if you were in the Bible?
I'm not in the Bible .
So these scriptures were random?
Not random, you read.
You said before You were in the Bible?
No direct reference. Again, I was taken out. My work was mentioned, but not where it came from, Me.

I wanted to check in on my friend Paul who I hadn't seen for a while. He has had a tough life and turned to drugs and alcohol to cope. He's one of the people I finally realized I couldn't save. I had to let him go. I'm sharing this with you to understand her perspective on how alcohol and drugs affect us.

Is my friend Paul still shielded?
No.
I thought You couldn't read him because he was shielded. I'm confused that You don't know his talents. When we come through the void You said You read aptitudes and offer assignments.
Use or lose.
I thought I had a resonance that was just me?
Resonance changes with drugs, age, mood, sex, etc.
Thought it was more.
Resonance is not aptitude. Resonance is where you are in space and time.
So he is not shielded right now?
He shielded himself from me before. Not any more. Not doing well. On drugs and doesn't care.

I remembered Her saying that when people do drugs or other mind altering substances our shields go down and anything can get in. It's interesting that even God can get in. When he is sober he can even shield himself from God. I'm sure She could go against his agency and look beyond the shield if She wanted to, but won't. Now he is an open book for all. Even the other side.

If I came from a traumatic childhood, did poorly in

272

school, and felt I had no talents, I wouldn't feel good either. No self love.

All too true. He has to be flat on his back in order to see up.

Is he doing hard drugs?

Yes.

Sad.

Yes.

I met someone who said they knew someone who talked to a relative from the other side who had died and that he met Jesus. When Jesus touched him he knew all truth. He realized his wife had never cheated on him and he had given her grief their whole life about it. When Jesus touched him he knew the truth that she hadn't cheated. He was trying to make amends from the other side to his wife who was still alive. I wanted to know if it is true that if Jesus touches us after we are dead, do we know all truth?

When we die and Jesus touches us, do we know all truth?

Can, case by case. If ready, and doesn't affect progress.

Can You eat whatever You want and not get fat?

Yes. Unless I want to. You can learn too.

Through meditation?

Close. Most humans don't know how.

How do I learn not to get fat?

You assume it's hard to get thin. When you know your true size it will happen. It is only a battle if you think it is.

You said we could do it too?

Not entirely, but don't be stupid. You can manifest resonance to being thin.

I went to the park, I didn't get clear answers today.

You're tired. My love not Dads.

What?

Hugs, not swats.

OK, that's right, Dad gives swats and Mom gives hugs.

December 5, 2012

Hi. Smudge new car. House shields are down. Fix salt around the house. Torg is sick. Rest. Not same sick as Ryxi, hers are hormones. Book wow, hair nice. Pictures tomorrow?

What genre is this book? New Age, Christian...?

All.

The Bible talks a lot about "the elect". What is your definition of Elect?

Common word. Even the Rothschilds use it.
Mine was "preordained servants". House of Israel.

But not all those You preordained are doing their calling.

Most fail their promise. Remember I told you: talents vs resonance.

So would you say that the elect are those being righteous? Or those doing their preordained callings?

Read the parable of grafting.

So elect are *all* those working for God, preordained or grafted in later.

True dat. People born with preset or resonance are under more condemnation if they stray.

From what you have said, I have done enough to be safe, but should do more?

You know the saying "boat in harbor?" (A boat in a harbor is safe, but that is not what it is made for.)

Yes. So why would You call someone later?

Resonance IS. No matter when. Resonance is not talent. Oneness is resonance.

Like New Agers would say, "That resonates."

Same vibes, man!

So I resonate with You, because You are a rebel bitch?

More than you know!

So you're a mean bitch in heaven?

Not mean, cool!

I'm still curious about how lesbians and homosexuals can become Gods.

Lesbians and homosexuals can achieve Our kingdom. This means they are Gods with all the abilities of other Celestial beings except for the ability to procreate and populate worlds with spirit children; this takes bonded male/female. They act primarily as Angels of Glory.

Is there anything you want me to keep out of the book? I'm worried about how people will react to the "bitch" word and information about gays and lesbians.

274

Yes, with explanation. Keep all in. Not be safe in ignorance.

I am taking out personal stuff.

Does it relate to the theme?

Yes, some are about relationships.

Then put it in. Like parables, examples help.

You said there is a node north of Sydney, Australia. You also said there will be a community like Safe Haven near Sydney, will the community be at the node?

Most nodes are not safe.

What about the Missouri node?

Not safe. It is a power center. The enemy knows them. Power centers are not safe. Look at Jerusalem.

I thought about how the Israelis and the Arabs fight over Jerusalem which was built on a node.

Not safe energetically, or due to people?

War for the Grid.

We still need to get to the Missouri node before the Solstice.

New plan. You send the onyx in the mail. Go later after the shift. The Mississippi river is guarded against you. It would draw Grid down to keep you safe. Do an end run. (She must mean like football, a sneaky play.) *Mail will not be expected.*

They would stop us if we flew over?

They expect people to come. Not expect mail. You could be in a plane, train, car, or walk. Still dangerous. Send with symbols you know. Better to be safe.

Why is it not safe to carry the onyx with symbols? How did Mac bring Pongust with symbols?

Mac was in general time and place. This is more specific, too dangerous. I am not ordering, just making a strong suggestion.

Yes, Mac didn't come during the solstice, and came to Salt Lake City which isn't a node. Made sense. The solstice is getting closer and the onyx hasn't arrived yet. Of course they would be watching!

Give us specifics?
     1. Send to the town next to the node. 2. Send to the postmaster as "general delivery". 3. Put name to pick up: Torg, Marcin, Ryxi.
Isn't Isis still in Missouri, couldn't we send it to her?
     Isis moved to New Mexico. 4. Make a medicine wheel. 5. Glue onyx on the back. 6. Wrap in paper with symbols. 7. Then foil. 8. No activator.
So it will be safe even without the activator?
     The activator makes the shield and breaks the shield.
When do we go after the shift?
     Maybe February or March.
That's when Marcin, who is a leader of another intentional community in our circle said we could come.
     Wow that!
Marcin is a nuclear physicist who gave up his job to create a sustainable intentional community. He came to stay with us five years ago and said the spirit was calling him to create his community in Missouri. He is Polish, go figure! We had randomly found his community in a magazine article recently. He put the community under the Open Mind Foundation which is also what Safe Haven Villages is under. We view them as a sister community. We've never been able to visit them. I was hoping we'd be able to visit during our trip for the solstice. When I looked him up online I saw he is only 70 miles from the node. No coincidence!!! We e-mailed him and he wrote back that it would be better to visit in February. I love synchronicity!

How's everything else?
     Give Rider a blessing before she goes to Mexico for Winter solstice. Have her take 4 colors of onyx from Adam's temple to leave at Palenque. (Mayan pyramid ruins) *Ask her to tap one to bring back. Send Ulu with her.*

(LeRoy came in.)
     Love you tender, love you. . .
     Joan of Arc is Ryxi's friend. You wear pants too.
Joan is a friend?!
     Yes. Why does it matter? You know many people more powerful than Joan!
Cool!

Mary too.

Which Mary?

>Both. You resonate.

Are they bitches too?

>Yes.

My photo shoot for the book cover was the next day and I had found a red tunic to wear. I wanted Her input.

Is Red a good color?

>Yes. Red lips, blood of Jesus. Wear heart necklace. Get a sword and wear a crown.

A crown and sword seemed really cheesy. Also, wouldn't it come across as being arrogant if I wore a crown?

I ended up doing two photo shoots because I didn't like the red tunic. I was prompted to go to a metaphysical store and found a neat ethereal looking green dress that matched the trees better. I decided to try a sword for fun. It actually worked! Warrior scribe Ryxi should have a sword, right? Didn't have the guts to try the crown. Maybe next time.

December 9, 2012

You said all your children are still there on your planet. You also said patriarchs are off being Gods in new universes, I'm confused?

>Many of the Patriarchs and Matriarchs are all over. But they live here.

Like other universes and dimensions?

>Just like Ryxi.

When you talk about Enoch's people, you say they were transfigured. What do you mean by transfigured?

>So you can see God and not die. Like Adam in the Garden, physical body changed, can't die.

What was in the apple that made him change.

>Not an apple.

Well, the fruit. Is this scientific?

>Long talk. Are you a biochemist? Everything is science.

Do you have more about transfiguration?

>Oh, yes. Heavenly mother 201.

Now that makes me wonder about your bodies.

Great bodies. Gravity yes. We made gravity.
Are you flesh and bone?
Of course. Torg hugged his dad. (Resurrected.)
Are translated beings physical, too?
Of course.
Can they float?
Yes. Not much reason. Usually they translocate. Like Stargate. We can float or fly if we want. Flying is much faster.
So you don't need to get into an airplane or flying saucer to get around the universe?
The universe is my saucer. I go anywhere I want when I want.
OK, do You get into something to get there?
Of course not.
Do You travel through worm holes?
Sort of. Humans not know yet.
Like a pin head?
Less. That's why making Gods is so important. The universe is big, all things seem small.
The book is taking a lot longer than we thought.
Keep going, and going, and going.
Love you.

I watched a Mayan video about being able to manifest more powerfully during solstice.

Are the Mayan's right, will the 2012 solstice be an exceptionally powerful time for manifestation?
Yes, you know this, peak. Don't wait for that, do now. Now is the time to prepare to meet God.
Kundalini spank is creation?
Sex, yes.

December 16, 2012
Great cover. Looks just like me.
Are you on the cover?
Yes, you are Me.

I have concerns about all these people going on shooting rampages. Are they being brainwashed. Why?
All humans are brainwashed. Schools, family, church, etc.
What about the Sandy Hook shooter?

I was getting to that! Yes, some people are brainwashed. Others are more dangerous. He only killed a couple dozen. Mussolini, Hitler, others killed whole generations. He was programmed for a high profile event, so are many other survivalists.

So the survivalists are evil too?

Many, yes.

So who programmed him?

SURVIVALISTS.

So it wasn't the governments, or the Illuminati?

No.

So the survivalists trained him on purpose?

Yes.

Why?

To bring things to a head.

Why did he kill his mom?

Hated her.

Why did he hate his mom?

Ruled too much.

So why did he target little kids?

Big splash.

Was it his idea? Did he meet with the survivalists and plan this.

No, he was tweaked to flip out. Could have chosen a shopping mall, post office. Others did.

How can survivalists be held accountable?

Can't.

But how did they program him?

Media, weapons.

But how?

Fight back at the MAN!

Was he planning this a long time?

Years. Your war of independence was not planned specifically. Just a lot of people on the edge.

Was there a trigger?

Holidays.

Why holidays?

Big splash.

He was troubled anyway, right?

Yes, many are.

Was this about his mom?

Not all.

How do You judge him?

Nurture.

279

Is he responsible?
> *Some. Nurture means teaching.*

So this was not a set up to lose our guns.
> *No, to keep the guns. Don't read too much into it.*

What about the guy at the Trax station in Salt Lake City who had a gun and was threatening people?
> *Too much stress.*

Was he a victim of mind control?
> *Yes, could have gotten help.*

What about the guy in Colorado?
> *You try to read too much. Much more subtle than that.*

Brainwashing just takes normal school.

They purposely do it subtly through normal school?
> *Yes. Some personality types are more susceptible. New subject!*

Leo: I got a download that I have been promoted to a higher level of a Shaman. What is this promotion?
> *It's good for you. Are you ready?*

What am I doing?
> *Power, teach, no more hiding.*

Why is so much going on in the spirit realm?
> *Big solstice, big mother of a solstice!*

Dec. 22, 2012
The 2012 Solstice

The big day we had been preparing for over the past several years was finally here! That morning I recalled the first Winter Solstice event I had attended in the beginning of my "awakening." It was 2007, I was in Tikal, Guatemala at a Mayan Temple with a Mayan Priestess named Aum Rak. We met in a circle on the grass at the foot of the temple where we started our ceremony. She told us we were Shamans called to assist Gaia to bring in the new Millennium and the return of Quetzalcoatl! She planned to have a solstice event every year up to 2012. It seemed so far away back then. Here it was, 2012, and I was spending it in my home rather than some exotic place to make sure we were "safe."

Around 11:30 our team of six men and six women showed up in time for Torg to take flight with the dragons and dance the new Grid system. We started off by telling him how much we appreciate him and how we needed him to come back. I wanted him to know he had a reason to come back.

He chose to be "in" the earth for his trip. He created sacred space in our home where he has been digging to create a basement. He wanted to be in the earth to help him ground and come back. We gathered around him, said prayers and sang songs. He then went into his trance state and left for the dance.

We then migrated to my backyard where I lead a ceremony in the sweat lodge. It didn't seem like very long until I heard the old familiar drumming out side the lodge that could only be Torg. I was glad to know he was back.

Dragon Dance
By Torg

For the solstice this year, the dragons performed their (our) dance for the third time. This time we were to dance to gentle the Gaia shift to the Millennial formation. She has been twisted over the ages by Continental Drift and other things to where the "soccer ball" pattern of the Grid is all but eradicated.

My team was chosen (as explained earlier) and the stage was set. Ryxi and the others celebrated a sweat lodge close by while I went to naked intelligence form for the dance. I had practiced so much, it was second nature by now.

The dance was boring! I had expected much more drama: music, fire arcs, others dancing! At least as much as when we Shamans overthrew the node at Madagascar. But I was assigned as a group leader with a number of other dragons (144?) under me to do very simple tasks like circling, or jumping up and down, or going in and out. After a while of this, we were simply dismissed and sent home. Without a word of thanks that I remember! It is now apparent that our job was essential to the overall activity, and that it was designed to not entice us to stay and be "light lost.' Still, I had been looking forward for years to repeating the original dance. Oh, well!

Torg

After the dance we talked to Mom and the elements and they said everything went well. Gaia stopped by and thanked us for a job well done. She said it would take another 10 years for her Ley lines to be completely reset and to expect more natural disasters in the process. We would still need to finish our world tour of the sacred sites to finalize the shift. I felt like this was a good place to end the book.

I can't wait to see what adventures HER new millennium has to offer! I pray that this book will inspire you to seek out your Heavenly Mother and allow her to heal your heart, guide your relationships and and make you laugh. Thanks for joining me on this ride and I look forward to sharing with you the next chapter of my adventure with Heavenly Mother in book 2. Until then,

Aloha,
Ryxi

Another Witness

My awakening with Divine Mother
By: Leroy Merworth

My spiritual path and relationship with God began in August of 2009, when I took a trip down to the Mohave desert in Arizona with my father and a friend of mine. My father was determined to find the lost treasure in the Superstition Mountains. We were hiking and soon found ourselves lost! Let me tell you, I was obviously not aware of what I was getting myself into. I had never been to Arizona. Getting lost was probably the most terrifying time in my life, I'll say! I prayed my heart out that we would find the way out. Thank God we were finally able to find a dried up river bed where we spent the night. Thank you God for my wonderful protective guides, because we never ran into any snakes, scorpions or anything deadly. I know, right?

The next morning we awakened to this bright light in the sky. Not like a helicopter, or a plane. It was silent, and the very bright light was right above us. Very close, too close. I asked it to move aside and it moved slowly aside.

Our instincts told us that the direction of the light was the way through to civilization. We followed the path on the old dried up river bed, heading north. Shortly after, we found power lines and a trail. On this trail we found a pile of rocks that looked as if someone purposely piled them up; like a marker of some kind. We saw human foot prints in the sand. It seemed as if someone was guiding us through. Then we saw a road! We were safe! Now, I look back and call it my crazy adventure! Maybe an accidental Vision Quest? It was definitely a baptism by fire awakening!

A year later, July of 2010, my father went back to the Superstition Mountains! I know, right? Why? Didn't he learn his lesson the first time? Was it Gold fever? The mountains kept calling, he couldn't resist. This time he never came back. The Superstition Mountains collected yet more lost souls obsessed with discovering her hidden treasures. I made my peace with him much later when he visited me in spirit explaining he had his reasons, a karmic debt that needed to be paid. Everything happens for a reason.

After much time and many search and rescue missions, we

283

finally found his body. The day we had his funeral, I had done my best to forgive him for going back.

For a year I was so depressed I almost committed suicide; something had to give. And then it shifted, the divine gave me a fresh start.

It happened one evening when, while praying, I finally surrendered to my higher power. I knew in my heart I was being heard. I questioned everything. The next day, my intuition was urging me to go to the Earthjam festival at Liberty park. There I met Ryxi and Torg at their Shamanism 101 booth. I knew I was being told to take Shamanism 101. I just knew this was my spiritual path, thanks Divine Mother.

I first met Heavenly Mother during the Shamanism 101 course, the day we were learning Matrix Energetics. I was playing with Ryxi's luminous energy field and just by tuning in, I saw rainbow luminous hair. I explained what I saw to Ryxi, and she said, "That must be Heavenly Mother. She is eccentric and likes to wear Her hair in many colors and styles." The more I got to know Her the more I learned; She is never without style! I wondered if She was expressing Her emotions and creativity. Or, was She just standing out in Her divine kingdom?

In time I have gotten to know Her, and have had many visits from Her. She is a busy woman; Her time is limited, but still finds time to visit me. As I got to know Her more I also learned that She can be outspoken and has a great sense of humor. She likes to make me blush and says things to shock me. I've heard She likes to surprise our divine Father, as well.

She encourages me to learn, to grow, and accept my path and calling. She says that the Earth is shifting and may be facing climate changes. She teaches me that we are entering Her new Millennium. She also says that Crystal children, spiritual warriors, care takers, healers and Shamans are going to defend the earth when war breaks out. The Divine has a design for all of us and many are being called, awakened to the awareness that our Earth Mother is needing good earth medicine healers. People are suffering. Many species of animals and plants are dying. Divine Mother is calling out. She calls us Earth Tied Power Walkers, or something like that. She encourages us to laugh, sing, and dance. It helps the power Grids in the earth. When I do I notice how it effects my own energy and health. Again, thank you Divine Mother.

We can all do our part in saving the earth. We can seek miracles when petitioning our Divine Mother. She always comes quick, fast, and in a hurry. Her sword can cut through all negativity if you ask it of Her.

By now I have fully accepted my role as a Shaman, an Earth Tied Grid Walker. I have grown to love the elements, my intelligences. And developed a relationship with my guides, angels, and ancestors that keeps me out of trouble! No hiking in hot deserts looking for treasure! I will always be grateful for all my father's love, and all of his sacrifices. Deep down in his daring attempts to look for, and find the lost Dutchman's gold, he was only trying to find his own inner riches. AKA: himself. Out of it all I have been given such tremendous spiritual gifts. I love you dad. Thank you angels, guides and divine beings for guiding me through it all. They may not come in ways we all imagine, as a feminine angelic being with wings, as I once believed. They come in many different forms. In fact, one evening I was tuning in to speak with my guides with my Shaman partner, and I saw a centaur standing near a tree. That was totally weird to me. I asked the centaur his name, he replied, "Sigmoore." I told him I thought centaur's didn't exist. He told me he was a divine messenger and he was sent to surround me with love, light, and protection. He was only helping me with my intrusive negative past life entity.

I asked divine Mother if centaurs are real. She said yes, but not in physical form. As a divine spirit, he works as a guide or an angel.

I've had many encounters with angels and learned they work to protect us from physical, mental, and spiritual depletion. They can come as animals, ancestors, loved ones that have crossed over and mythological creatures. Anyone can be someone else's angel. The old saying, "One small act of kindness can create one big ripple effect," is how it really works. We are creating energetic ripples that are unseen to most people's physical eyes. Our guides are only serving to protect, guide, and teach us for a divine purpose. I love them, they are all my relations.

Thank you, Divine Heavenly Mother; I love Your colorful personality. You have changed my life with Your divine unconditional love. I love you. Thank you for choosing me to serve others and our Earth Mother and be a part of the divine Shamanic team. You are always praising us. We praise You. Thank You for inspiring us to laugh, sing, and dance. Thanks for coming into our hearts. Our beloved Heavenly Mother. I pray that each and everyone of us come into an awakening with Your divine team. Enjoy the ride! God bless. Amen.

By Leroy Merworth.

INDEX

www.ingramcontent.com/pod-product-compliance
Lightning Source LLC
Chambersburg PA
CBHW060251100426
42742CB00011B/1710